ENERGY SCIENCE, ENGINEERING AND TECHNOLOGY

GRID ENERGY STORAGE

BENEFITS AND CHALLENGES TO MODERNIZING THE ELECTRIC GRID

ENERGY SCIENCE, ENGINEERING AND TECHNOLOGY

Additional books in this series can be found on Nova's website under the Series tab.

Additional E-books in this series can be found on Nova's website under the E-book tab.

ENERGY SCIENCE, ENGINEERING AND TECHNOLOGY

GRID ENERGY STORAGE

BENEFITS AND CHALLENGES TO MODERNIZING THE ELECTRIC GRID

JONATAN JANKO
EDITOR

nova publishers
New York

For permission to use material from this book please contact us:
Telephone 631-231-7269; Fax 631-231-8175
Web Site: http://www.novapublishers.com

Library of Congress Cataloging-in-Publication Data

ISBN: 978-1-63117-368-4

Published by Nova Science Publishers, Inc. † New York

CONTENTS

PREFACE

Energy storage technologies (such as pumped hydro, compressed air energy storage, various types of batteries, flywheels, electrochemical capacitors, etc.) provide for multiple applications that include energy management, backup power, load leveling, frequency regulation, voltage support, and grid stabilization. Importantly, not every type of storage is suitable for every type of application, motivating the need for a portfolio strategy for energy storage technology. There are four challenges related to the widespread deployment of energy storage. The challenges are cost competitive energy storage technologies (including manufacturing and grid integration), validated reliability and safety, equitable regulatory environment, and industry acceptance. This book sets out potential options to improve energy storage. It also presents a number of specific actions that could help maintain both scientific advancements and a pipeline of project deployments.

Chapter 1 – Modernizing the electric system will help the nation meet the challenge of handling projected energy needs—including addressing climate change by integrating more energy from renewable sources and enhancing efficiency from non-renewable energy processes. Advances to the electric grid must maintain a robust and resilient electricity delivery system, and energy storage can play a significant role in meeting these challenges by improving the operating capabilities of the grid, lowering cost and ensuring high reliability, as well as deferring and reducing infrastructure investments. Finally, energy storage can be instrumental for emergency preparedness because of its ability to provide backup power as well as grid stabilization services.

At present, the U.S. has about 24.6GW (approx. 2.3% of total electric production capacity) of grid storage, 95% of which is pumped storage hydro.

Europe and Japan have notably higher fractions of grid storage. Pursuit of a clean energy future is motivating significantly increased storage development efforts in Europe and Asia, as well as the U.S.

Energy storage technologies—such as pumped hydro, compressed air energy storage, various types of batteries, flywheels, electrochemical capacitors, etc., provide for multiple applications: energy management, backup power, load leveling, frequency regulation, voltage support, and grid stabilization. Importantly, not every type of storage is suitable for every type of application, motivating the need for a portfolio strategy for energy storage technology.

Chapter 2 – Energy storage is one of several potentially important enabling technologies supporting large- scale deployment of renewable energy, particularly variable renewables such as solar photovoltaics (PV) and wind. Although energy storage does not produce energy—in fact, it is a net consumer due to efficiency losses—it does potentially allow greater use of variable renewables by shifting energy from periods of low demand to periods of high demand, which reduces curtailment and eases integration challenges. Energy storage can also provide a variety of high value services such as firm capacity and multiple ancillary services.

Energy storage is used in electric grids in the United States and worldwide. It is dominated by pumped-storage hydropower (PSH), with about 20 GW[164] deployed in the United States and more than 127 GW deployed worldwide. In the United States, PSH was built largely in response to market conditions in the 1970s, including high oil and natural gas prices, regulatory restrictions on plants burning oil and gas, dependence on low-efficiency steam plants for peaking power, and anticipated "build-out" of a largely inflexible nuclear fleet. In addition to PSH, a single, 110-MW compressed air energy storage (CAES) facility has been constructed in the United States.

In: Grid Energy Storage
Editor: Jonatan Janko
ISBN: 978-1-63117-368-4

Chapter 1

GRID ENERGY STORAGE*

U.S. Department of Energy

EXECUTIVE SUMMARY

Modernizing the electric system will help the nation meet the challenge of handling projected energy needs—including addressing climate change by integrating more energy from renewable sources and enhancing efficiency from non-renewable energy processes. Advances to the electric grid must maintain a robust and resilient electricity delivery system, and energy storage can play a significant role in meeting these challenges by improving the operating capabilities of the grid, lowering cost and ensuring high reliability, as well as deferring and reducing infrastructure investments. Finally, energy storage can be instrumental for emergency preparedness because of its ability to provide backup power as well as grid stabilization services.

At present, the U.S. has about 24.6GW (approx. 2.3% of total electric production capacity) of grid storage, 95% of which is pumped storage hydro.[1] Europe and Japan have notably higher fractions of grid storage. Pursuit of a clean energy future is motivating significantly increased storage development efforts in Europe and Asia, as well as the U.S.

Energy storage technologies—such as pumped hydro, compressed air energy storage, various types of batteries, flywheels, electrochemical

* This is an edited, reformatted and augmented version of the U.S. Department of Energy, dated December 2013.

capacitors, etc., provide for multiple applications: energy management, backup power, load leveling, frequency regulation, voltage support, and grid stabilization. Importantly, not every type of storage is suitable for every type of application, motivating the need for a portfolio strategy for energy storage technology.

There are four *challenges* related to the widespread deployment of energy storage: cost competitive energy storage technologies (including manufacturing and grid integration), validated reliability & safety, equitable regulatory environment, and industry acceptance. Issues that are being explored in this paper focus on reducing system costs through targeted application of science and engineering research and development for new storage concepts, materials, components and systems (including manufacturability and standardization). Developers should consider technical risk mitigation, for controlling the uncertainties at the early stage of deployment so that cost estimates and operational practices can develop based upon well-grounded and fully understood data. Ongoing research and development, from fundamental science of energy storage mechanisms to the early stage development of platform technologies should also be considered in support of these challenges. Industrial standards for grid storage are in their infancy. Industry acceptance could also gain ground when we reduce the uncertainty surrounding how storage technology is used, and monetized, at scale. Ultimately, it will be the experience and real-world use of storage that will provide the confidence and desire to expand installed storage.

The expansion of the electricity system can be accelerated by the widespread deployment of energy storage, since storage can be a critical component of grid stability and resiliency. ***The future for energy storage in the U.S. should address the following issues: energy storage technologies should be cost competitive (unsubsidized) with other technologies providing similar services; energy storage should be recognized for its value in providing multiple benefits simultaneously; and ultimately, storage technology should seamlessly integrate with existing systems and sub-systems leading to its ubiquitous deployment*.**

In reviewing the barriers and challenges, and the future for energy storage, a strategy that would address these issues should comprise three broad outcome-oriented goals:

1. Energy storage should be a broadly deployable asset for enhancing renewable penetration – specifically to enable storage deployment at high levels of new renewable generation

2. Energy storage should be available to industry and regulators as an effective option to resolve issues of grid resiliency and reliability
3. Energy storage should be a well-accepted contributor to realization of smart-grid benefits – specifically enabling confident deployment of electric transportation and optimal utilization of demand-side assets.

To realize these outcomes, the principal challenges to focus on are:

- **Cost competitive energy storage technology** - Achievement of this goal requires attention to factors such as life-cycle cost and performance (round-trip efficiency, energy density, cycle life, capacity fade, etc.) for energy storage technology as deployed. It is expected that early deployments will be in high value applications, but that long term success requires both cost reduction and the capacity to realize revenue for all grid services storage provides.
- **Validated reliability and safety** - Validation of the safety, reliability, and performance of energy storage is essential for user confidence.
- **Equitable regulatory environment** – Value propositions for grid storage depend on reducing institutional and regulatory hurdles to levels comparable with those of other grid resources.
- **Industry acceptance** – Industry adoption requires that they have confidence storage will deploy as expected, and deliver as predicted and promised.

DOE is addressing these challenges in the following ways:

Challenge/Goal	Strategy Summary
Cost competitive energy storage technology	• Targeted scientific investigation of fundamental materials, transport processes, and phenomena enabling discovery of new or enhanced storage technologies with increased performance • Materials and systems engineering research to resolve key technology cost and performance challenges of known and emerging storage technologies (including manufacturing) • Seeded technology innovation of new storage concepts • Development of storage technology cost models to guide R&D and assist innovators • Resolution of grid benefits of energy storage to guide technology development and facilitate market penetration

(Continued)

Challenge/Goal	Strategy Summary
Validated reliability and safety	• R&D programs focused on degradation and failure mechanisms and their mitigation, and accelerated life testing • Development of standard testing protocols and independent testing of prototypic storage devices under accepted utility use cases • Track, document, and make available performance of installed storage systems
Equitable Regulatory Environment	• Collaborative public-private sector characterization and evaluation of grid benefits of storage • Exploration of technology-neutral mechanisms for monetizing grid services provided by storage • Development of industry and regulatory agency-accepted standards for siting, grid integration, procurement, and performance evaluation
Industry acceptance	• Collaborative, co-funded field trials and demonstrations enabling accumulation of experience and evaluation of performance – especially for facilitating renewable integration and enhanced grid resilience • Adaptation of industry-accepted planning and operational tools to accommodate energy storage • Development of storage system design tools for multiple grid services

1.0. INTRODUCTION

Modernizing the electric grid will help the nation meet the challenge of handling projected energy needs—including addressing climate change by relying on more energy from renewable sources—in the coming decades, while maintaining a robust and resilient electricity delivery system. By some estimates, the United States will need somewhere between 4 and 5 tera kilowatt-hours of electricity annually by 2050.[2] Those planning and implementing grid expansion to meet this increased electric load face growing challenges in balancing economic and commercial viability, resiliency, cyber-security, and impacts to carbon emissions and environmental sustainability. Energy storage systems (ESS) will play a significant role in meeting these challenges by improving the operating capabilities of the grid as well as mitigating infrastructure investments. ESS can address issues with the timing, transmission, and dispatch of electricity, while also regulating the quality and reliability of the power generated by traditional and variable sources of power. ESS can also contribute to emergency preparedness. Modernizing the grid will require a substantial deployment of energy storage. In the past few years, the urgency of energy storage requirements has become a greater, more pressing issue that is expected to continue growing over the next decade:

- California enacted a law in October 2010 ***requiring*** the California Public Utilities Commission (CPUC) to establish appropriate 2015 and 2020 ***energy storage procurement targets for California load serving entities, if cost effective and commercially viable by October 2013*** (AB 2514). In February 2013, the CPUC determined that Southern California Edison must procure 50 MW of energy storage capacity by 2021 in Los Angeles area. Additionally, in June 2013, the CPUC proposed storage procurement targets and mechanisms totaling 1,325 MW of storage. Other States are looking to the example that California is setting, and Congress has introduced two bills that establish incentives for storage deployment.[3]
- The increasing penetration of renewable energy on the grid ***to meet renewable portfolio standards*** (RPS) may be linked with greater deployment of energy storage. Storage can "smooth" the delivery of power generated from wind and solar technologies, in effect, increasing the value of renewable power. Additionally, when energy storage is used with ***distributed generation***, it can improve the reliability of those assets by providing power-conditioning value, and enables increased renewable penetration to help contribute to meeting state RPS.
- Energy storage is already near commercial viability in augmenting power management and frequency regulation techniques. Large flywheel installations and power monitoring software have combined to make flywheel installation useful in ensuring that intermittent sources and variable load demands maintain a 60 Hz frequency, storage could be an alternative method of providing spinning reserve or curtailment which could ***improve the efficiency of infrastructure and reduce greenhouse gas emissions*** caused by wasteful excess capacity and lowered heat-rates associated with excessive plant cycling.

- Energy storage can reduce the need for major new transmission grid construction upgrades as well as ***augment the performance of existing transmission and distribution assets.*** DOE estimates that 70% of transmission lines are 25 years or older, 70% of power transformers are 25 years or older, and 60% of circuit breakers are more than 30 years old.[4] Extending the capability of the transmission grid—for example by pre-positioning storage on the load side of transmission constraint points—makes the grid more secure, reliable, and responsive. Additionally, distributed storage can reduce line-congestion and line-loss by moving electricity at off-peak times, reducing the need for overall generation during peak times. By reducing peak loading (and overloading) of transmission and distribution lines, storage can extend the life of existing infrastructure.
- Moreover, as the nation moves towards the ***electrification of the transportation sector***, energy storage for vehicles, and the integration of energy between vehicles and the grid, will be critical. The focus on storage is not only for the deployment of batteries in vehicles, but also for potential second-life applications for electric vehicle (EV) batteries. For example, Project Plug-IN, a large scale public/private EV initiative based in Indianapolis, involving Duke Energy, is exploring the best customer use for stationary applications in homes, neighborhoods, and commercial buildings. This pilot project is being used to help validate the business models for future commercialization of storage technologies.
- Energy storage will also play a significant role in emergency preparedness and increasing overall grid resilience. An August 2013 White House report,[5] written in conjunction with the Office of Electricity Delivery & Energy Reliability, details the integral role that energy storage will play in enhancing grid resilience and robustness related to weather outages and other potential disruptions.[6]
- Energy storage is poised to grow dramatically, requiring large investment in manufacturing capacity and jobs. According to an Information Handling Services, Cambridge Energy Research Associates (IHS CERA) report, ***the energy storage business could grow from $200 million in 2012 to a $19 billion industry by 2017***.[7]

The Department of Energy serves a vital role in resolving major challenges that are hampering widespread deployment of grid energy storage. Teaming with industry, State and municipal governments, academia, and other

Federal agencies, DOE supports the discovery of new technologies to improve cost and performance of grid energy storage, spur technology innovation and incorporation into improved storage products, remove unnecessary barriers to deployment, and facilitate the establishment of industry-wide standards to ease widespread adoption of storage. These activities can help to ***catalyze the timely, material, and efficient transformation of the nation's energy system and secure U.S. leadership in clean energy technologies.*** DOE's 2011 Strategic Plan has identified a number of targeted outcomes in support of this goal, the most relevant to this mission includes reducing energy storage costs 30% by 2015 and supporting the integration of Plug-in Hybrid Electric and Battery Electric Vehicles as they shift load profiles[8].

Storage technology can help contribute to overall system reliability as large quantities of wind, solar, and other renewable energy sources continue to be added to the nation's generation assets, furthering the goals of reducing greenhouse gas emissions and increasing energy security. Additionally, storage technology will be an instrumental tool in managing grid reliability and resiliency by regulating variable generation and improving microgrid and smart-grid functionality. For micro- and smart-grid technologies, storage can provide redundancy options in areas with limited transmission capacity, transmission disruptions, or volatile demand and supply profiles.

The Department's electric energy storage program can create economic opportunities, as well. A strong storage market will foster a robust manufacturing base of advanced electric energy storage devices in the U.S., and this capability can be leveraged for export opportunities in the robust foreign market for storage. Further, by enabling more efficient adoption of renewable energy sources in the US, storage can help promote US energy independence and reduce carbon emissions.

Overview of this Report

This report sets out potential options to improve energy storage. It also presents a number of specific actions that could help maintain both scientific advancements and a pipeline of project deployments. *This report does not address new policy actions, nor does it specify budgets and resources for future activities.*

Section 2.0 of this report describes the present state of energy storage in the US, as well as international projects that could serve as a near-term template for US investment and growth. Section 3.0 describes the present state of technology for energy storage, including the applications and opportunities for each technology type. Section 4.0 discusses the barriers and challenges to widespread adoption of grid-storage techniques, as well as other concerns that will need to be addressed. Sections 5.0 and 6.0 highlight ideas on how to promote and advance energy storage over the next three to five years, ranging from promoting basic research to promoting and analyzing present and future grid-storage markets. Section 7.0 discusses goals related to technology developments while section 8.0 discusses goals related to analysis. Finally section 9.0 addresses standardization and DOE's ongoing activities. The appendices detail storage R&D programs at relevant DOE offices and several Federal agencies and provides a listing of American Recovery and Reinvestment Act of 2009 (ARRA) funded energy storage projects.

2.0. State of Energy Storage in US and Abroad

An interactive database[9] created and maintained by DOE provides a snapshot of the extent and range of energy storage systems deployments worldwide. As of August 2013, the database reported 202 storage system deployments in the US with a cumulative operational capability of 24.6 GW, with a mix of storage technologies including pumped hydro, various types of batteries, and flywheels.[10] The contribution of each technology to the overall operational capability is shown in Figure 1. At 95%, pumped hydro clearly dominates due to its larger unit sizes and longer history as the technology of choice for energy storage by the electric utility sector. Other technologies such as compressed air energy storage (CAES), thermal energy storage, batteries, and flywheels constitute the remaining 5% of overall storage capability.

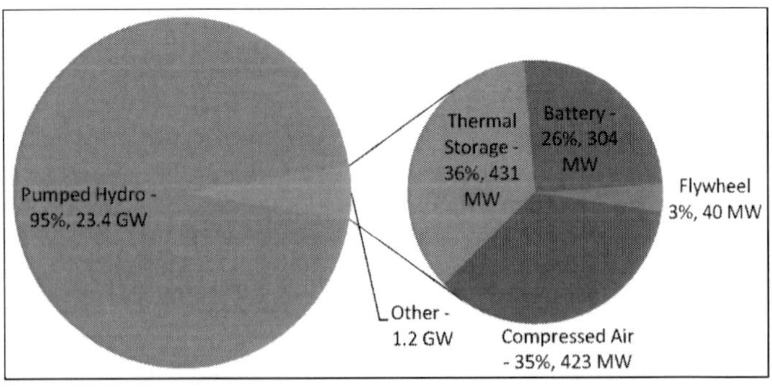

Figure 1. Rated Power of US Grid Storage projects (includes announced projects).

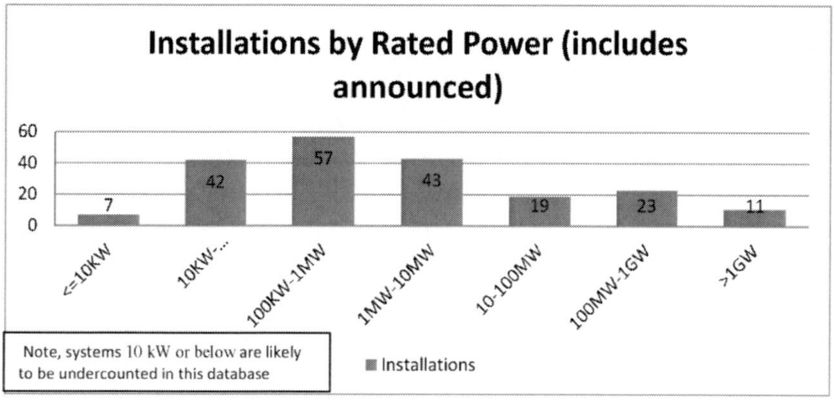

Figure 2. Number of US installations, grouped by capacity.

Similarly, Figure 2 shows the wide range of system sizes that have been deployed. The rated power of the various projects ranges from small, residential scale (7 projects are listed as 10 kW or below—this is a reporting artifact, as there are likely many small systems not in the database) to large, utility scale systems of 1 MW or more.[11]

Energy storage systems and the services they provide can be used in regulated and deregulated markets. However, for energy storage technologies used on the grid, regulatory policies and rules provide the framework for the business case and economics of storage systems. Other incentives, such as tax structures and asset depreciation rates significantly affect the economics for storage projects. All the electrical grid-connected storage services, market opportunities, cost-recovery methods, cost-effectiveness criteria, incentives, and rebates are governed by a well-established regulatory oversight. The Federal Energy Regulatory Commission (FERC) regulates interstate transactions, while State entities such as Public Utility Commissions (PUCs) regulate utility management, operations, electricity rate structures, and capacity acquisition within their State's jurisdiction. Additionally, in some regions Independent System Operators (ISOs) provide oversight of transmission and generation. This multi-level oversight impacts the growth of the storage industry because policies can create or inhibit market opportunities for electricity storage and may determine how, and if, they will be compensated.

New policies are being implemented at the State level, being discussed and rolled out at the national level, and previous investments are coming to fruition and can shape future investment.[12] As an example of the influence of policy structure on the adoption of storage, FERC Order 755 helps structure payments and set contracts for frequency regulation, and is changing the market for frequency-regulation applications. PJM was the first Regional Transmission Organization (RTO) to adopt Order 755, and the results have significantly improved the commercial viability of frequency regulation. Further, the frequency regulation market will likely continue to mature, as several other RTOs have or are scheduled to adopt Order 755. For example, Midcontinent Independent System Operator (MISO) also adopted the order at the end of 2012, while the California and New York Independent System Operators (CAISO & NYISO) adopted the mandate in mid-2013 and the Independent System Operator - New England (ISONE) will begin following the Order in January 2014.[13] Additionally, Congress continues to debate two bills[14] that would help codify the cost structure for storage-related subsidies and partnership taxation structures for investment in storage and storage-related activities.

In addition to national developments, California, Texas, New York, Hawaii, and Washington have all proposed significant policies on storage. California has enacted laws that make energy storage more viable from a cost and regulatory perspective and give the California Public Utilities Commission (CPUC) the power to mandate certain regional penetration levels of storage. The CPUC recently mandated that 50 MW of storage be installed in the Los Angeles Basin by 2020, as well as a top-line mandate of 1.3 GW of storage for the entire state.[15] The Texas legislature has enacted SB 943 that classifies energy storage technologies alongside generation equipment, and the Public Utility Commission of Texas adopted key aspects of the bill as well as clarified rules, requirements, and definitions for energy storage[16]. In 2010, New York State established NY Battery and Energy Storage Technology Consortium (NY-BEST),[17] a public-private partnership that researches storage technology and manufacturing, aids energy storage organizations as well as potential stakeholders, and advocates for policies and programs that could improve energy storage. Additionally, Washington State enacted two laws[18] related to energy storage: the first enables qualifying utilities to credit energy storage output of renewable sourced energy at 2.5 times the normal value; the second requires electric utilities to include energy storage in all integrated resource plans.

On the national level, several projects that were funded under ARRA through the Smart- Grid Demonstration Grant program are coming online in 2013, and their performance has the potential to guide future investment decisions and policy initiatives. In total, an estimated 59 MW of storage capacity is scheduled to come online in 2013, accounting for 7 of the 16 ARRA-funded projects. In addition, hydrogen fuel cells for backup power are being used in more than 800 units associated with telecom towers in the U.S., as a result of ARRA funding.[19]

Internationally, Japan has pursued the development and deployment of energy storage to balance the variability of load on its nuclear power plants. After completing an initial phase of building pumped hydro storage plants, Japan pursued development of other storage technologies. Its most prominent accomplishment was the commercial development of high temperature sodium-sulfur batteries through a sustained R&D program that spanned two decades.[20] Today, Japan-based NGK is the only source of sodium-sulfur batteries and as of 2012, NGK had over 450 MW of sodium sulfur storage systems installed.[21]

China and India are also pursuing energy storage programs to support the rapid growth in their electric energy needs. Energy storage could serve many grid needs in both China and India to bridge the gap between available generation and customer loads during system peaks and as a distributed resource on the customer-side of the meter. In one example, India is aggressively pursuing energy storage as a secure power resource for more than 300,000 telecom towers, and announced a $40 million contract in July 2013 for Li-ion battery energy storage systems to meet that need. This example has the potential to demonstrate telecom towers as a "first market" for storage technologies developed and manufactured in the U.S.[22]

Table 1 describes some of the country-specific highlights of international grid storage.

Technology Overview

Storage systems can be designed with a broad portfolio of technologies such as pumped hydro, compressed air energy storage CAES, a large family of batteries, flywheels, and superconducting magnetic energy storage (SMES). Each technology has its own performance characteristics that makes it optimally suitable for certain grid services and less so for other grid applications. This ability of a storage system to match performance to different grid requirements also allows the same storage system to provide multiple services. This gives storage systems a greater degree of operational flexibility that cannot be matched by other grid resources, such as combustion turbines or a diesel generator. The ability of a single storage system to meet multiple requirements also makes it feasible to capture more than one value stream, when possible, to justify its investment. While the categorization of "deployed," "demonstrated," and "early stage," is often blurred, and changes over time, Figure 3 groups technologies based on their present degree of maturity.[24]

This portfolio of electricity storage technologies can provide a range of services to the electric grid and can be positioned around their power and energy relationship. Established large-scale technologies, such as CAES and pumped hydro, are capable of discharge times in tens of hours and with high module sizes that reach 1,000 MW. Pumped hydroelectric energy storage is a large, mature, and commercial utility-scale technology currently used at many locations in the United States and around the world. Pumped hydro currently employs off-peak electricity to pump water from a reservoir up to another reservoir at a higher elevation. When electricity is needed, water is released from the upper reservoir through a hydroelectric turbine into the lower reservoir to generate electricity. New capabilities of pumped hydro, through the use of variable speed pumping, is opening up the potential for the provision of additional services that may be used to assist in the integration of variable generation sources. Projects may be practically sized up to 4,000 MW and operate at about 76%–85% efficiency, depending on design. Pumped hydro plants have long lives, on the order of 50-60 years. As a general rule, a reservoir one kilometer in diameter, 25 meters deep, and having an average head of 200 meters would hold enough water to generate 10,000 MWh. CAES systems are not as "mature" as pumped hydro, but are similar in their use as they store energy in the form of pressurized air, usually in underground caverns. However, both CAES and pumped hydro have very specific geographic requirements making their installation site-dependent.

For example a CAES plant can enhance the grid to allow successful integration of significant amounts of renewable resources while enhancing the transmission system and providing grid stability during intermittent operations. It will also provide ancillary services such as regulation, capacity reserve, and reactive power for voltage support with the size and flexibility that will enable large amounts of energy to be stored and discharged for use to maintain and improve the grid system reliability and relieving transmission congestion. Energy storage, such as CAES enhances the grid by making the grid more efficient, which will assist in achieving the full potential of renewables and will provide an industry model for a grid-enabled diversified energy portfolio.

Table 1. International Landscape of Grid Storage

Country	Storage Targets [23]	Projects	Other Issues	Technology & Applications
Italy	75 MW	• 51 MW of Storage Commissioned by 2015 • Additional 24 MW funded	• Italy has substantial renewables capacity relative to grid size, and the grid is currently struggling with reliability issues; additional renewables capacity will only exacerbate this problem	• 35 MW to be Sodium-Sulfur Batteries for long-duration discharge • Additional capacity is focused on reliability issues and frequency regulation
Japan	30 MW	• Approved 30 MW of Lithium-ion battery installations	• Potential decommissioning of nuclear fleet • Large installation of intermittent sources - est. 9.4 GW of solar PV installed in 2013 alone • Several isolated grids with insufficient transmission infrastructure during peak demand periods	• Primarily Lithium ion batteries • Recently increased regulatory approved storage devices from 31 to 55
South Korea	154 MW	• 54 MW lithium-ion batteries • 100 MW CAES	• Significant regulatory/performance issues with nuclear fleet	• Reliability & UPS
Germany	$260m for grid storage	• $172m already apportioned to announced projects	• Decommissioning entire nuclear fleet; Large (and expanding) intermittent renewable generation capabilities • Over 160 energy storage pilot projects • Awaiting information on energy storage mandates	• Hydrogen; CAES & Geological; Frequency Regulation
Canada	-	• Announced 1st frequency regulation plant	-	-
UK	-	• 6 MW multi-use battery	• Other small R&D and Demonstration projects	• Battery will perform both load shifting and frequency regulation applications

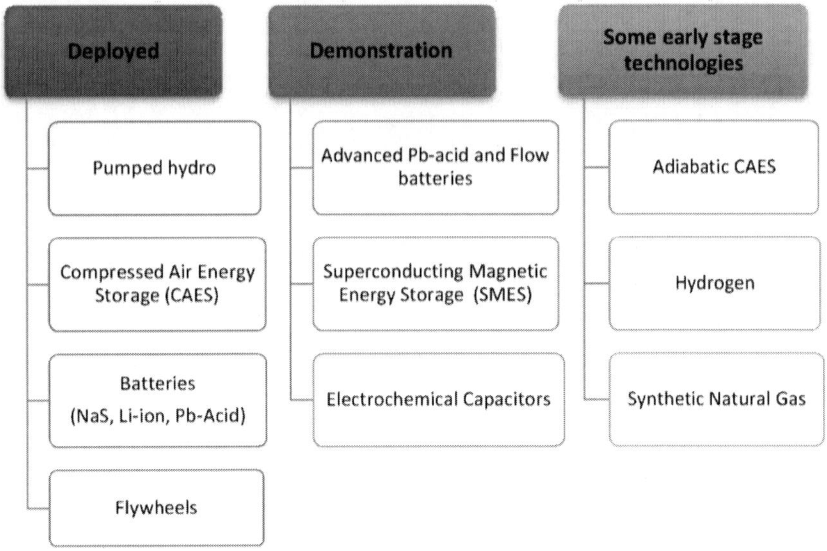

Figure 3. Maturity of electricity storage technologies.

Table 2. Installations of Batteries

Battery Type	Number of 1MW+ Deployments	Largest Installation
Lithium Ion	15	40
Sodium Sulfur	11	4*
Lead Acid	9	36

*There are two 4MW deployments.

In contrast to the capabilities of these two technologies, various electrochemical batteries and flywheels are positioned around lower power and shorter discharge times, ranging from a few seconds to six hours, and these technologies can generally be built without specific geographical features at the site.

There are several different electrochemical battery technologies that are currently available for commercial applications. These technologies have been successfully deployed in both distributed and centralized applications in various sizes. However, they have not yet realized widespread deployment due to challenges in energy density, power performance, lifetime, charging capabilities, safety, and system cost. The more robust technologies include lithium-ion (Li-ion), sodium sulfur (NaS), and lead acid batteries, including lead carbon batteries. Li-ion batteries tend to be best suited for relatively short discharges (under two hours) and do not handle deep-discharges well, so these batteries are more suited to power-management operations such as frequency regulation or as an uninterruptible power source (UPS). NaS batteries are somewhat behind Li-ion battery technology in terms of energy and power, but they can maintain longer discharges (four to eight hours) and may be more suitable for load leveling and price arbitrage operations. Lead acid batteries, a mature technology with good battery life, are relatively cheap; however, the low energy density and short cycle time are challenges to large-scale deployment. Also, there are other novel chemistries being developed such as sodium-ion.

Flywheels are currently commercially deployed primarily for frequency regulation. Flywheel plants take in electricity and convert it into spinning discs, which can be sped up or slowed down to rapidly shift energy to or from the grid, which ensures steady power (60 Hz) supplied to the grid.

Flow batteries were invented by US utilities specifically to provide MW-scale storage capacity beyond the geographic constraints of hydroelectric facilities. Significant US industry and DOE investment over the past 40 years has led to a mature understanding of the advantages and limitations of available chemistries, as well as more recent breakthroughs in performance and thermal tolerance. However, due to lack of MW-scale field history, flow batteries have not gained substantial commercial traction in the US, with various flow-battery technologies still in the demonstration phase, and the largest single operational system at 0.6 MW.[25] However, recently we are seeing flow batteries projects launch oversees with systems up to 5MW in size and a total deployed capacity of 20MW. China and Japan are currently funding over $200MM in flow battery projects and Europe is following suit with numerous smaller projects. The interest in flow batteries stems from several potential advantages over traditional batteries, primarily the liquid suspension and separation of the chemical components that allow for full charge utilization with a high number of discharge cycles and extremely long unit life. Flow batteries have faced obstacles related to their low energy density and integrated design requirements that make it difficult to compete at sub-MW scale. With recent advances in these areas, flow batteries may be commercially deployable in the US within the next few years if MW-scale projects similar to current ARRA projects succeed.

Another technology in the demonstration/applied research phase is superconductive magnetic energy storage (SMES). Each unit employs a superconducting coil, a power conditioning system, and a cooling system. The cooling system chills the coil below the superconducting transition temperature, so that electrical currents flow without resistance or loss of energy. Energy is stored inductively in the DC magnetic field of a solenoid, as long as the temperature remains sufficiently low. Most SMES technologies currently have a high cycle-life and power density, but low energy density and high cost that make them best suited for supplying short bursts of electricity into the energy system. Superconductors currently have the highest round-trip efficiency of any storage device, though they are costly to manufacture and maintain and they have only a limited number of small demonstrations.

Electrochemical capacitors (EC) technology stores direct electrical charge in the material, rather than converting the charge to another form, such as chemical energy in batteries or magnetic field energy in SMES; this makes the storage process reversible, efficient, and fast. As such, EC can be useful in power-quality applications such as frequency regulation, and voltage stabilization. The devices may have longer useful lives since there is little breakdown in the capacitors ability to store energy electrostatically. Currently, electrochemical capacitors can store significantly more energy than dielectric and electrolytic capacitors; however, EC technology is still cost prohibitive.[26]

Thermochemical energy storage is an emerging technology that uses reversible chemical reactions to store heating or cooling capacity in chemical compounds. The promise of thermochemical storage is the tremendous energy densities that it can achieve over most other storage types, ranging from 5 to 20 times greater than conventional storage. Due to its relatively high energy density potential, a significant research and development effort is currently being focused on this type of thermal energy storage, though deployments are limited.

Hydrogen systems, as with the other storage technologies, require careful analysis to fully capture the value stream. Multiple components such as electrolyzers, fuel cells, or hydrogen oxygen turbines coupled with storage, either underground in geologic formations or above ground in hydrogen tanks, can be used in grid systems. Hydrogen can also allow for the decoupling of electricity production and storage resulting in flexible operation. While round trip energy efficiencies might be at a level of 40%, this relatively low efficiency is balanced by energy storage potential that may last days, to weeks, or longer.

Table 3 summarizes the state of most energy storage technologies.

Table 3. Technology Types Source: Advancing Energy Storage

Technology	Primary Application	What we know currently	Challenges
CAES	• Energy management • Backup and seasonal Reserves • Renewable integration	• Better ramp rates than gas turbine plants • Established technology in operation since the 1970's	• Geographically limited • Lower efficiency due to roundtrip conversion • Slower response time than flywheels or batteries • Environmental impact
Pumped Hydro	• Energy management • Backup and seasonal reserves • Regulation service also available through variable speed pumps	• Developed and mature technology • Very high ramp rate • Currently most cost effective form of storage	• Geographically limited • Plant site • Environmental impacts • High overall project cost
Fly wheels	• Load leveling • Frequency regulation • Peak shaving and off peak storage • Transient stability	• Modular technology • Proven growth potential to utility scale • Long cycle life • High peak power without overheating concerns • Rapid response • High round trip energy efficiency	• Rotor tensile strength limitations • Limited energy storage time due to high frictional losses
Advanced Lead-Acid	• Load leveling and regulation • Grid stabilization	• Mature battery technology • Low cost • High recycled content • Good battery life	• Limited depth of discharge • Low energy density • Large footprint • Electrode corrosion limits useful life

Table 3. (Continued)

Technology	Primary Application	What we know currently	Challenges
NaS	• Power quality • Congestion relief • Renewable source integration	• High energy density • Long discharge cycles • Fast response • Long life • Good scaling potential	• Operating Temperature required between 250° and 300° C • Liquid containment issues (corrosion and brittle glass seals)
Li-ion	• Power quality • Frequency regulation	• High energy densities • Good cycle life • High charge/ discharge efficiency	• High production cost - scalability • Extremely sensitive to over temperature, overcharge and internal pressure buildup • Intolerance to deep discharges
Flow Batteries	• Ramping • Peak Shaving • Time Shifting • Frequency regulation • Power quality	• Ability to perform high number of discharge cycles • Lower charge/discharge efficiencies • Very long life	• Developing technology, not mature for commercial scale development • Complicated design • Lower energy density
SMES	• Power quality • Frequency regulation	• Highest round-trip efficiency from discharge	• Low energy density • Material and manufacturing cost prohibitive
Electrochemical Capacitors	• Power quality • Frequency regulation	• Very long life • Highly reversible and fast discharge	• Currently cost prohibitive
Thermochemical Energy Storage	• Load leveling and regulation • Grid stabilization	• Extremely high energy densities	• Currently cost prohibitive

Table 4. Electric Grid Energy Storage Services

Bulk Energy Services	
	Electric Energy Time-Shift (Arbitrage)
	Electric Supply Capacity
Ancillary Services	
	Regulation
	Spinning, Non-Spinning and Supplemental Reserves
	Voltage Support
	Black Start
	Other Related Uses

Transmission Infrastructure Services	
	Transmission Upgrade Deferral
	Transmission Congestion Relief
Distribution Infrastructure Services	
	Distribution Upgrade Deferral
	Voltage Support
Customer Energy Management Services	
	Power Quality
	Power Reliability
	Retail Electric Energy Time-Shift
	Demand Charge Management

3.0. GRID SCALE ENERGY STORAGE APPLICATIONS

Until the mid-1980s energy storage was viewed by the electric utilities as a means to time shift energy produced by coal and nuclear units during off-peak hours to displace energy that would be produced from other more expensive fuels during on-peak periods. Several factors, including environmental concerns in building large pumped hydro plants and the emergence of other storage technologies using batteries and flywheels, introduced the viability of using storage to provide other grid services.[27]

The 2013 edition of the DOE/EPRI Electricity Storage Handbook describes eighteen services and applications in five umbrella groups, as listed in Table 4. The services and applications identified in this table show that energy storage can be used to support generation, transmission, and distribution, as well as customer-side-of-the-meter needs of the grid. This section describes some of the functions most commercially viable and relevant to the near-term future of the grid. [28]

Recognizing energy storage can have multiple services within the grid allows it to capture multiple benefit streams to offset system costs. The flexibility of storage can be leveraged to provide multiple or stacked services, or use cases, with a single storage system that captures several revenue streams to achieve economic viability. How these services are stacked depends on the location of the system within the grid and the storage technology used. However, due to regulatory and operating constraints, stacking services is a process that requires careful planning and should be considered on a case-by-case basis. The following are brief discussions of some applications of grid energy storage:

Electric Energy Time-shift (Arbitrage)

Electric energy time-shift involves purchasing inexpensive electric energy, available during periods when prices or system marginal costs are low, to charge the storage system so that the stored energy can be used or sold at a later time when the price or costs are high. Alternatively, storage can provide similar time-shift duty by storing excess energy production, which would otherwise be curtailed, from renewable sources such as wind or photovoltaic (PV). The functional operation of the storage system is similar in both cases, and they are treated interchangeably in this discussion.

Ancillary Services: Regulation & Frequency Response

Regulation is one of the ancillary services for which storage is especially well suited. Regulation involves managing interchange flows with other control areas to match closely the scheduled interchange flows and momentary variations in demand within the control area. The primary reasons for including regulation in the power system are to maintain the grid frequency and to comply with the North American Electric Reliability Council's (NERC's) Real Power Balancing Control Performance (BAL001) and Disturbance Control Performance (BAL002) Standards, which are mandatory reliability standards approved by FERC

Regulation is used to reconcile momentary differences caused by fluctuations in generation and loads. Regulation is used for damping of that difference. Consider the example shown in Figure 4: the load demand line shows numerous fluctuations depicting the imbalance between generation and load without regulation. The thicker line in the plot shows a smoother system response after damping of those fluctuations with regulation.

Generating units that are online and ready to increase or decrease power as needed are used for regulation and their output is increased when there is a momentary shortfall of generation to provide up regulation. Conversely, regulation resources' output is reduced to provide down regulation when there is a momentary excess of generation. An important consideration in this case is that large thermal base-load generation units in regulation incur some wear and tear when they provide variable power needed for regulation duty.

Frequency response is very similar to regulation, except it reacts to system needs in even shorter time periods of less than a minute to seconds when there is a sudden loss of a generation unit or a transmission line. As shown in Figure 5,[29] various generator response actions are needed to counteract this sudden imbalance between load and generation to maintain the system frequency and stability of the grid. The first response within the initial seconds is the primary frequency control response of the governor action on the generation units to increase their power output as shown in the lower portion of the figure. This is followed by the longer duration secondary frequency control response by the AGC that spans the half a minute to several minutes shown by the dotted line in the lower portion of Figure 5. It is important to note that the rate at which the frequency decays after the triggering event – loss of generator or transmission – is directly proportional to the aggregate inertia within the grid at that instant. The rotating mass of large generators and/or the aggregate mass of many smaller generators collectively determines this inertia.

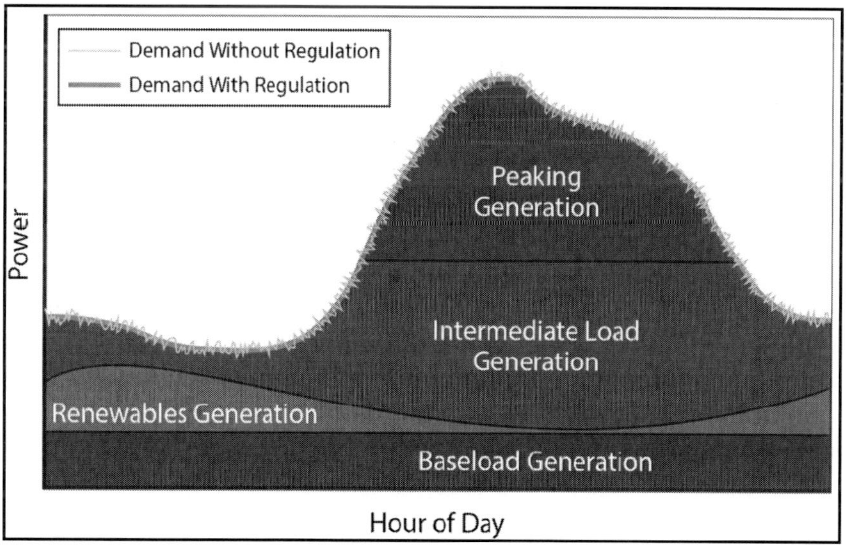

Figure 4. System Load Without and With Regulation.

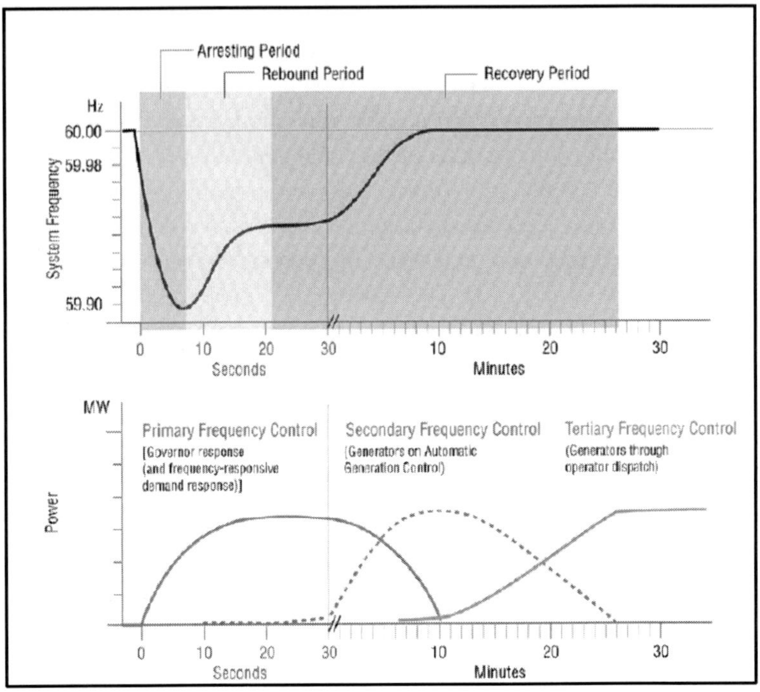

Figure 5. The Sequential Actions of Primary, Secondary, and Tertiary Frequency Controls Following the Sudden Loss of Generation and Their Impacts on System Frequency.

The combined effect of inertia and the governor actions determines the rate of frequency decay and recovery shown in the arresting and rebound periods in the upper portion of Figure 5. This is also the window of time in which the fast-acting response of flywheel and battery storage systems excels in stabilizing the frequency. The presence of fast-acting storage assures a smoother transition from the upset period to normal operation if the grid frequency is within its normal range.

Spinning, Non-Spinning, and Supplemental Reserves

Operation of an electric grid requires reserve capacity that can be called upon when some portion of the normal electric supply resources become unavailable unexpectedly.

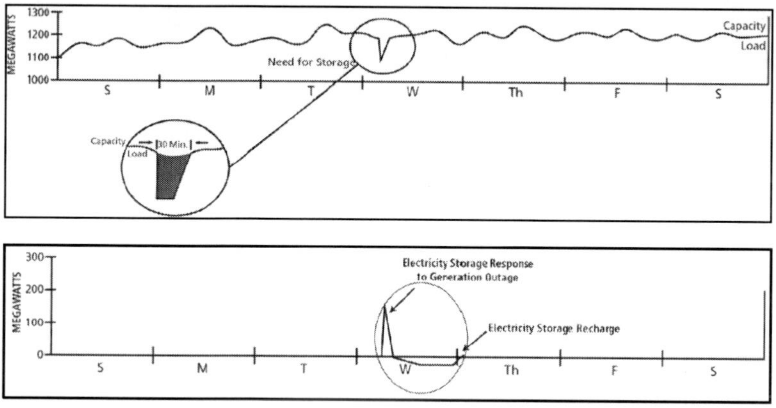

Figure 6. Storage for Reserve Capacity.

Generally, reserves are at least as large as the single largest resource (e.g., the single largest generation unit) serving the system and reserve capacity is equivalent to 15% to 20% of the normal electric supply capacity. NERC and FERC define reserves based on different operating conditions. There are three generic types of reserves: spinning or synchronized reserves[30] that can respond within 10 seconds to 10 minutes to service frequency issues, or generation or transmission outages; non-spinning or non- synchronized reserves[31] that can respond within 10 minutes for use as uninterruptible and/or curtailable loads; and supplemental reserves that can pick up load within an hour to back up any disruption to spinning and non-spinning reserves. Importantly for storage, generation resources used as reserve capacity must be online and operational (i.e., at part load). Unlike generation, in almost all circumstances, storage used for reserve capacity does not discharge at all; it just has to be ready and available to discharge when needed.

Reserve capacity resources must receive and respond to appropriate control signals. Figure 6 shows how storage responds to spinning reserve requirements. The upper plot shows a loss of generation and the lower plot shows the immediate response with a 30- minute discharge to provide the reserve capacity until other generation is brought online.

Load Following/Ramping Support for Renewables

Electricity storage is eminently suitable for damping the variability of wind and PV systems and is being widely used in this application. Technically, the operating requirements for a storage system in this application are the same as those needed for a storage system to respond to a rapidly or randomly fluctuating load profile. Most renewable applications with a need for storage will specify a maximum expected up- and down-ramp rate in MW/minute and the time duration of the ramp.[32] This design guidance for the storage system is applicable for load following and renewable ramp support.

Load following is characterized by power output that generally changes as frequently as every several minutes.[33] The output changes in response to the changing balance between electric supply and load within a specific region or area. Output variation is a response to changes in system frequency, timeline loading, or the relation of these to each other that occurs as needed to maintain the scheduled system frequency and/or established interchange with other areas within predetermined limits. A simple depiction of load following was shown in Figure 4 above.

Storage can alleviate some of the cycling of power plants through frequency regulation and other short-term power management techniques. Energy storage technologies can perform better than the existing system as recognized by FERC Order 755. They can potentially achieve efficiencies of 70 to 95 percent while operating at partial capacity with lower efficiency penalties while still maintaining near-instantaneous response times.

Distribution Upgrade Deferral and Voltage Support

Distribution upgrade deferral involves using storage to delay or avoid investments that would otherwise be necessary to maintain adequate distribution capacity to serve all load requirements. The upgrade deferral could be a replacement of an aging or over-stressed existing distribution transformer at a substation or re-conducting distribution lines with heavier wire.

When a transformer is replaced with a new, larger transformer, its size is selected to accommodate future load growth over the next 15- to 20-year planning horizon. Thus a large portion of this investment is underutilized for most of the new equipment's life. The upgrade of the transformer can be deferred by using a storage system to offload it during peak periods, thus extending its operational life by several years. Notably, for most nodes within a distribution system, the highest loads occur on just a few days per year, for just a few hours per year. Often, the highest annual load occurs on one specific day with a peak somewhat higher than any other day. One important implication is that storage used for this application can provide significant benefits with limited or no need to discharge.[34] Further, if the storage system is containerized, then it can be physically moved to other substations where it can continue to defer similar upgrade decision points. Additionally, deferring investment decisions reduces the burden of forecasting future increases in load demands; by delaying infrastructure and capacity investments, storage can bridge load demand in areas that see high demand growth or obviate investment where demand growth is below forecast.

Also, a storage system that is used for upgrade deferral could simultaneously provide voltage support on the distribution lines. Utilities regulate voltage within specified limits[35] by tap changing regulators at the distribution substation and by switching capacitors to follow load changes. This is especially important on long, radial lines where a large load such as an arc welder or a residential PV system may be causing unacceptable voltage excursions on neighboring customers. These voltage fluctuations can be effectively damped with minimal draw of real power from the storage system.

Customer Side of the Meter Storage

Increasingly, deployment of new technologies on the customer side of the meter is changing the electrical nature of customer requirements. And the evolution of the "smart grid" is enabling customers to shape their requirements to improve their own utilization of electricity, while also contributing to improved grid reliability, performance, and economics. These developments create an opportunity for customer storage to play an increasing role in grid services.

Energy storage has been used by customers for many years to achieve either improved reliability or economic benefits. Uninterruptable power supplies have long been used by customers requiring high reliability to provide short-term power to bridge the period between an outage and the start of backup generation. In some cases electrical storage provides backup power as well. More recently, customer systems combining PV and storage have increased because of their ability to improve customer energy economics. As PV systems drop in price, customers will increasingly opt to deploy PV-storage systems. In addition, thermal storage has been utilized for many years to reduce electrical usage during peak periods. And the deployment of electric vehicles (EV) is another form of customer storage of electricity. Co-optimized charging can be conducted in a manner that supports improved grid reliability and economics. Studies are underway now to examine the potential for using

EVs as power sources for the grid, essentially in the same manner that utility storage would be. Once retired (nominally when their energy capacity reaches 80% of their initial value), EV batteries can see secondary use, repackaged for providing stationary grid storage (currently being demonstrated by DOE).

In general, energy storage on the customer side of the meter is configured and optimized to meet customer service needs. The value proposition for storage is therefore customer specific, but from a grid perspective has primarily been driven by time-of-use rates and demand charges. A residential electric storage unit might be nominally 1.5- 5kW, and 3- 20kWh, while a commercial electric storage system would typically range from 10's of kW to multi-megawatt systems. Commercial customers will also acquire electric energy storage for power quality and reliability purposes. As utilities seek to increase the utilization of load to assist grid operation, principally through new market products and incentives for different functions (e.g. regulation and load following services), the value proposition for customer energy storage will evolve and is expected to increase. Similarly, where incentive structures for renewable deployment is joined with energy storage, the internal return on investment for the combined system can be significantly higher than for storage alone[36]. Beyond direct customer benefits (such as comfort or equipment sizing and performance), and absent special incentive structures, energy storage behind the meter will have to compete against utility-sited energy storage for grid services. Hence, cost targets described in Section 5 should serve as a threshold for behind the meter storage when used solely for grid services. In fact, combining grid service benefits with customer service benefits should raise the threshold of affordability for customer storage.

Table 5 on the next page summarizes many of these key applications by energy storage technology.

4.0. SUMMARY OF KEY BARRIERS

There are four barriers that should be explored to promote the widespread deployment of energy storage:

- Cost competitive energy storage systems
- Validated performance and safety
- Equitable regulatory environment
- Industry acceptance

Table 5. Applications by Technology Type

Application	Description	CAES	Pumped Hydro	Flywheels	Lead-Acid	NaS	Li-ion	Flow Batteries
Off-to-on peak intermittent shifting and firming	Charge at the site of off peak renewable and/ or intermittent energy sources; discharge energy into the grid during on peak periods	◑	◑	○	●	●	●	●
On-peak intermittent energy smoothing and shaping	Charge/discharge seconds to minutes to smooth intermittent generation and/or charge/discharge minutes to hours to shape energy profile	○	○	◑	●	●	●	●
Ancillary service provision	Provide ancillary service capacity in day ahead markets and respond to ISO signaling in real time	◑	◑	◑	◑	◑	◑	◑
Black start provision	Unit sits fully charged, discharging when black start capability is required	◑	◑	○	●	●	●	●
Transmission infrastructure	Use an energy storage device to defer upgrades in transmission	○	○	○	●	●	●	●
Distribution infrastructure	Use an energy storage device to defer upgrades in distribution	○	○	○	●	●	●	●
Transportable distribution- level outage mitigation	Use a transportable storage unit to provide supplemental power to end users during outages due to short term distribution overload situations	○	○	○	◑	●	●	●
Peak load shifting downstream of distribution system	Charge device during off peak downstream of the distribution system (below secondary transformer); discharge during 2-4 hour daily peek	○	○	○	●	●	●	●

Table 5. (Continued)

Application	Description	CAES	Pumped Hydro	Flywheels	Lead-Acid	NaS	Li-ion	Flow Batteries
Intermittent distributed generation integration	Charge/Discharge device to balance local energy use with generation. Sited between the distributed and generation and distribution grid to defer otherwise necessary distribution infrastructure upgrades	○	○	○	●	◑	◑	◑
End-user time-of-use rate optimization	Charge device when retail TOU prices are low and discharge when prices are high	◑	◑	○	◑	◑	◑	◑
Uninterrupti-ble power supply	End user deploys energy storage to improve power quality and /or provide back up power during outages	○	○	◑	●	●	●	●
Micro grid formation	Energy storage is deployed in conjunction with local generation to separate from the grid, creating an islanded micro-grid	○	○	○	●	●	●	●

Definite suitability for application ●; Possible use for application ◑ ; Unsuitable for application ○.

Cost Competitive Energy Storage Systems

The total cost of storage systems, including all the subsystem components, installation, and integration costs need to be cost competitive with other non-storage options available to electric utilities. While there is a strong focus on reducing the cost of the "storage" components, such as batteries or the flywheel, the storage component still constitutes only 30% to 40% of the total system cost, thus the focus needs to be on the entire system.

Additionally, there is a concurrent need to quantify the "value" of storage in the various services it provides to the grid, individually and in multiple or "stacked" services, where a single storage system has the potential to capture several revenue streams to achieve economic viability. This is important now and as the cost of storage systems decline to economically attractive levels.

Validated Performance and Safety

The process for evaluating and reporting the performance of existing storage systems on a unified basis needs to be created. This combined with industry accepted codes and standards to specify desired performance parameters for each storage service, will lead to a wider acceptance of energy storage systems. For example, there is significant uncertainty over the usable life of batteries[37] and the length of time that a storage installation can generate revenue; both of these issues directly impact investment calculations. According to stakeholder input, a fuller understanding of the true life of batteries through demonstrations and accelerated testing could help remove this barrier, since predicting reliability through improved testing is important in supporting commercialization.

The operational safety of large storage systems is a concern and will be a barrier in their deployment in urban areas or in proximity of other grid resources such as substations. Design practices that incorporate safety standards and safety testing procedures for the different storage technologies need to be developed and codified.

Equitable Regulatory Environment

Currently, a consistent pricing or market plan for providing grid storage does not exist and the uncertainty surrounding use-case economics inhibits investment. Without an established revenue generation model for storage operators, the case for investment will remain muted. While there have been demonstrations in areas such as frequency regulation, there are still enough revenue uncertainties in other applications to dissuade investment.

Industry Acceptance

There is also significant uncertainty about how storage technology will be used in practice and how new storage technologies will perform over time in applications. Currently, systems operators have limited experience using deployed storage resources; stakeholder input suggests that development of algorithms to employ storage technology effectively and profitably could encourage investments. Similarly, today's utility planning, transmission and distribution design tools do not have the capability to analyze energy storage as an option on a consistent basis. Integrating storage into the planning tools that are currently used by industry (rather than developing stand-alone tools) could boost storage technologies.

These four challenges were addressed during a recent DOE-sponsored workshop/webinar on energy storage, where industry and academic participants also noted:

- Balance of system costs are critically important and further analysis and research is needed to reduce them
- There should be a focus on manufacturability and reliability, and companies that have produced a viable application should receive support for manufacturing improvements
- Coordination is required to inform standard setting organizations regarding uniformity in product performance and interfaces
- There should be a thorough analysis of completed ARRA projects; successful ARRA projects should be encouraged to continue with commercial deployment of their technology, and they should be part of a regular pathway to continue projects leading to commercialization-ready prototypes.

5.0. ENERGY STORAGE STRATEGIC GOALS

The vision for the electricity system of the future includes a significant scale-up of clean energy and energy efficiency that balances environmental and energy goals with impacts on consumer costs and economic productivity. The adoption of technology for the two-way flow of energy and communications would open up access to information, participation, choice, and empower consumers with options from using electric vehicles to producing and selling electricity. The future grid will provide a coordinated balance between centralized, decentralized and automated control including interactions with microgrids, and while it becomes increasingly accessible to new technologies and innovation, it will remain reliable and secure against cyber and physical threats, and be resilient to disruptions and outages.

Given the important role energy storage will have in future electricity systems, a strategy for storage depends on focusing on how ***that energy storage technologies should be cost competitive (unsubsidized) with other technologies providing similar services; that energy storage should be recognized for its value in providing multiple benefits simultaneously; and that ultimately, storage technology should seamlessly integrate with existing systems and sub- systems leading to its ubiquitous deployment***.

In reviewing the barriers and challenges, and the future for energy storage, a ***strategy*** that would address these issues should comprise three broad outcome-oriented goals:

1. Energy storage should be a broadly deployable asset for enhancing renewable penetration– specifically to enable storage deployment at high levels of new renewable generation
2. Energy storage should be available to industry and regulators as an effective option to resolve issues of grid resiliency and reliability
3. Energy storage should be a well-accepted contributor to realization of smart-grid benefits – specifically enabling confident deployment of electric transportation and optimal utilization of demand-side assets.

To realize these outcomes, the principal challenges should be addressed as described in Section 4.0 above. To that end:

- **Cost competitive energy storage technology** - Achievement of this goal requires attention to factors such as life-cycle cost and performance (round-trip efficiency, energy density, cycle life, capacity fade, etc.) for energy storage systems as deployed. It is expected that early deployments will be in high value applications, but that long term success requires both cost reduction and the capacity to realize revenue for all grid services storage provides.

- **Validated reliability and safety** - Validation of the safety, reliability, and performance of energy storage is essential for user, investor and insurer confidence.
- **Equitable regulatory environment** – Value propositions for grid storage depend on reducing institutional and regulatory hurdles to levels comparable with those of other grid resources.
- **Industry acceptance** – Industry adoption requires that they have confidence storage will deploy as expected, perform and deliver as predicted and promised.

DOE has conducted workshops[38] with industry, and have developed the following cost and performance targets for near-term and long-term storage technology development for the grid:

Near-Term

- Demonstrate AC energy storage systems involving redox flow batteries, sodium-based batteries, lead-carbon batteries, lithium-ion batteries and other technologies to meet the following electric grid performance and cost targets:[39]
  - System capital cost: under $250/kWh
  - Levelized cost: under 20 ¢/kWh/cycle
  - System efficiency: over 75%
  - Cycle life: more than 4,000 cycles
- Develop and optimize power technologies to meet AC energy storage system capital cost targets under $1,750/kW

Long-Term[40]

- Research and develop new technologies based on advanced materials and chemistries to meet the following AC energy storage system targets:
  - System capital cost: under $150/kWh

- Levelized cost: under 10 ¢/kWh/cycle (i.e., economically scalable without subsidies)
- System efficiency: over 80%
- Cycle life: more than 5,000 cycles

- Develop and optimize power technologies to meet AC energy storage system capital cost targets under $1,250/kW
- For Concentrated Solar Power (CSP)-storage systems:
    - System capital cost: under $15/kWh
    - System efficiency: 95%
    - Cycle life: 10,000 cycles

Table 6. Strategy Summary for DOE Energy Storage

Challenge/Goal	Strategy Summary
Cost competitive AC energy storage systems	• Targeted scientific investigation of fundamental materials, transport processes, and phenomena enabling discovery of new or enhanced storage technologies • Materials and systems engineering research to resolve key technology and system cost and performance challenges of known and emerging storage technologies (including manufacturing) • Seeded technology innovation of new storage concepts • Development of storage technology cost models to guide R&D and assist innovators • Resolution of grid benefits of energy storage to guide technology development and facilitate market penetration
Validated reliability and safety	• Enhancement of R&D programs focused on degradation and failure mechanisms and their mitigation, and accelerated life testing • Development of standard testing protocols and independent testing of prototypic storage devices under accepted utility use cases • Track, document, and make available performance of installed storage systems
Equitable Regulatory Environment	• Collaborative public-private sector characterization and evaluation of grid benefits of storage • Exploration of technology-neutral mechanisms for monetizing grid services provided by storage • Development of industry and regulatory agency-accepted standards for siting, grid integration, procurement, and performance evaluation

Table 6. (Continued)

Challenge/Goal	Strategy Summary
Industry acceptance	• Collaborative, co-funded field trials and demonstrations enabling accumulation of experience and evaluation of performance – especially for facilitating renewable integration and enhanced grid resilience • Adaptation of industry-accepted planning and operational tools to accommodate energy storage • Development of storage system design tools for multiple grid services

An excellent discussion of the current attributes of many grid storage technologies are provided in the Electric Storage Handbook, reference 39, which was prepared in conjunction with the Electric Power Research Institute and the National Rural Electrical Cooperative Association. The information provided in the Handbook offers insight on the attributes of specific storage technologies. No technology currently meets all metrics; however, each technology has attributes that allow it to approach some metrics. In some markets, current technologies, which do not satisfy all the metrics, are already commercially viable. DOE's strategy to achieve these goals is outlined in Table 6.

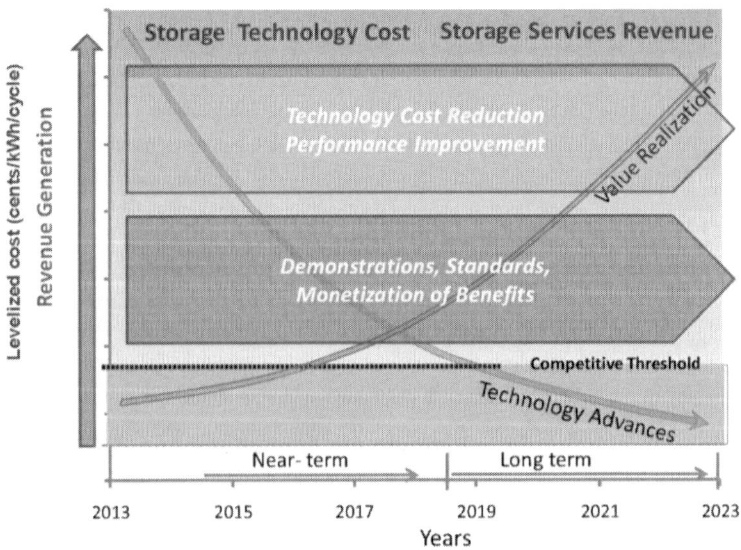

Figure 7. Storage Technology Cost.

The general strategy can be summarized as follows: Technology costs are driven down by focused research across a broad array of technologies. Storage provides a suite of application benefits, yet those benefits are not fully monetized. Therefore, efforts also focus on enabling realization of those benefits, their value, and enabling those who deploy storage to receive appropriate financial compensation. Even if financially prudent, deployment won't occur unless users (utilities and customers) and utility regulators have confidence in technology safety, reliability and performance. Hence a strategy that pursues collaborative field demonstrations; modification of design; planning; and operational tools; standardization; and equitable regulatory treatment; serves to gain confidence and reduce barriers to storage deployment. Figure 7 depicts thc outcome of this general strategy.

RPS requirements imply some 20% renewable generation by 2020. A PNNL study shows that this corresponds to some 18.6 GW of intra-hour balancing required for grid stability. Energy storage, demand response, and fast gas turbines can cover this need. A contribution of 5GW for energy storage is a reasonable lower bound. In addition, estimates by Pike Research set new deployment of energy storage at 14GW worldwide by 2022. This indicates that ***deployment of 5GW of new grid storage by 2025 is an achievable objective***[41].

6.0. Implementation of Goals

The issues, strategy, and goals described above are a framework to guide the deployment of energy storage.

DOE sponsors research, development, and demonstrations across multiple offices. The Office of Science/Basic Energy Science, the Office of Energy Efficiency and Renewable Energy, the Advanced Research Projects Agency for Energy, and the Office of Electricity Delivery and Energy Reliability, as well as the office of the Under Secretary for Science & Energy, all actively participate in energy storage programs. See Appendix A for brief descriptions of related activities in each office.

The general roles of these offices relative to the risk and technology readiness are illustrated in Figure 8 and Table 7. In energy storage, as in many other technologies, as risk is reduced and technology matures, the private sector and those public sectors active in deployment take on greater roles and responsibilities. DOE's role changes from that of providing scientific and technology advances during the early stages of technology development to one of independent analyst, convener, and facilitator addressing common issues affecting technology adoption.

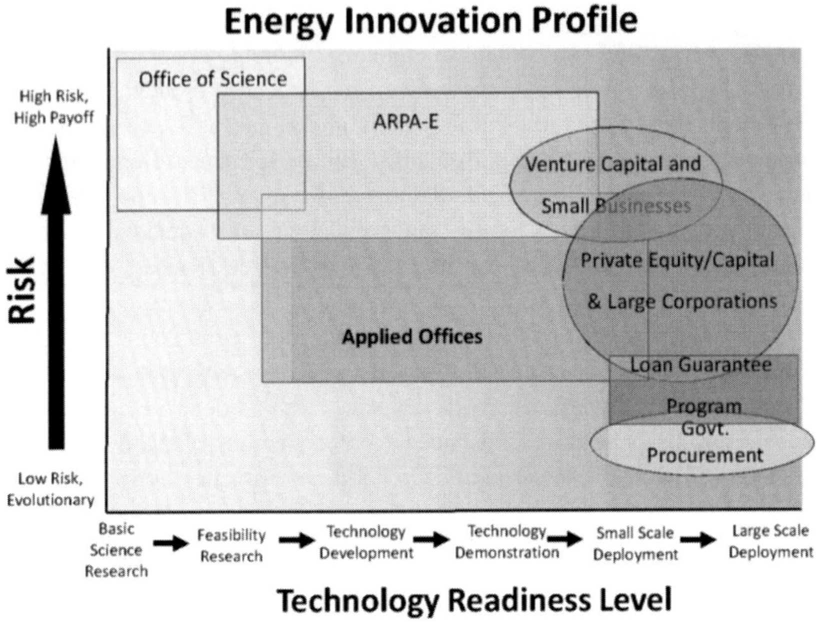

Figure 8. Role of DOE Offices in Technology Development, Maturation and Commercialization.

Table 7. Role of DOE Offices in Grid Energy Storage

OFFICE	Role
OE	Energy storage technology research, development, modeling, demonstration and control technologies that can increase the flexibility and resilience of the electrical grid), from the generator to the consumer, thereby allowing for grid integration of a greater diversity of technologies. This includes microgrids with storage, including for emergency preparedness.
EERE	R&D and demonstrations of on-site energy storage technologies that enable penetration of EERE technologies (generation, efficiency, or transportation) into the current system (grid). Additionally, removal of siting and permitting challenges faced by pumped storage deployment and the evaluation of how variable speed pumped storage can provide ancillary services and add to system flexibility.
ARPA-E	High risk R&D to prove & prototype disruptive new energy storage technologies.

OFFICE	Role
SC-BES	Fundamental research to (i) design and develop novel materials and concepts and (ii) probe physical and chemical phenomena associated with electrical energy storage.
LPO	Debt financing of commercial energy projects which include innovative storage technologies.

Strategy Implementation

The DOE strategic goals may be pursued by a coordinated suite of efforts, summarized below in Table 8. While a specific DOE activity may be primarily to accomplish one goal, many activities contribute to achieving multiple goals. For instance, fundamental materials research helps achieve lower cost and higher performance storage with long service life, but also will contribute to the establishment of standards for accelerated cycle life testing. Similarly demonstration projects, as acknowledge by EPRI and industry broadly, support improvements in system reliability, performance, value discovery, grid integration standards, and refinement of planning and operational tools.

Table 8. Specific Activities in Support of the Energy Storage Strategy

Cost Competitive Energy Storage Technology	• Advanced redox-flow battery chemistry and component development to utilize lower cost membranes, electrodes, bi-polar plates; increase energy (electrolytc composition) and power density (ion flux across the membrane); develop non-aqueous redox-flow – bench scale test of potential \$250/kW system • Bench demonstration of low temperature sodium (Na) batteries with efficiency of 90%– metal halide, and Na-ion systems • Demonstration at relevant scale of 2nd use automotive, and safer, longer-lived stationary Li-ion systems • Bench demonstration of multi-cell Mg-ion batteries for stationary applications • Develop and test high performance nano-material-based flywheel components • Advanced SiC, GaN, and AlN-based power converters for storage applications • Develop and test new capacitor materials and structures Use directed advanced research for • *Flywheels*: Thick cross section carbon fiber composite structure formation; magnetically levitated system enabling technology research, new nano-structured flywheel and magnet materials

Table 8. (Continued)

	• *CAES*: Proof-of-scientific- concept isothermal CAES research; • *Electrochemical*: novel low cost, high cycle life anode, cathode, electrolyte and separator materials and structure research • *Flow-Batteries*: High current density, low cost power modules, long cycle-life low-cost electrodes, membranes and catalysts; alkaline exchange membrane electrodes and multi-functional power/energy electrodes; semi-solid flow-able anolyte/catholytes; nanostructured electrode assemblies • *Superconductors*: low cost high-temperature superconducting materials; high-field coil configurations • *Capacitors*: High surface area, nanostructured ultracapacitors, low cost, safe, and stable electrolytes/solvents • *Power electronics*: Novel inverter/converter topologies; integrated passive components; high voltage wide bandgap semiconductor epitaxial materials • *Batteries*: Low-cost/high energy density batteries; rechargeable metal-air chemistries; in-situ sensors and control technologies; model predictive cell control algorithms; moldable energy storage structures; multi-functional storage chemistries
	• Benefit/cost analysis grid integration of storage for grid resilience, emergency response, renewable deployment, and improved asset utilization • Storage cost models (including manufacturing) to guide R&D and for industry use • Development of design tools for optimally serving multiple applications • Baseline techno-economic modeling of advanced research impacts; value-proposition development for emerging technology research results; first-market analysis for subsequent technology insertion; partnership formation for direct private sector or other governmental hand-offs
Validated Reliability & Safety	• Independent testing of prototype storage materials, components and devices in both lab and field systems • Forensic investigation of degraded storage from materials to systems • Degradation, failure, and safety processes/mechanisms characterization and models • Validation of accelerated life-cycle testing protocols • Documentation of field demonstrations regarding performance • Technology specific testing to support hand-offs to governmental and private sector partners, following initial de-risking of advanced research concepts • User facility for validation and testing of system performance
Equitable Regulatory Environment	• Documentation of federal, state and local policies affecting storage deployment • Review of IRP and similar regional, state and community analytic processes affecting storage development and deployment

	• Exploration of alternative policies that may affect technology attributes and deployment • Support development of consensus based codes and standards for performance, safety, packaging, cycle life, control and grid integration of storage
	• Maintenance of publicly available information on storage technology and attributes affecting its deployment • Dissemination of comprehensive information on storage technology status, experience (e.g., ARRA projects), and realizable contributions to grid resilience, emergency response, renewable deployment, and asset utilization • Provide best practices for installation and use of energy storage to regulators, policy makers, and industry
Industry/ Stakeholder Acceptance	• Conduct analyses and develop tools assessing the beneficial role of storage in cost-effectively achieving higher levels of renewable deployment Provide independent analytic support to public-private sector studies and field trials/demonstrations characterizing the benefits and costs of storage to facilitate renewable deployment
	• Collaborate with industry on enhancement of production cost models, transient event, and other grid simulation/analysis tools to accurately incorporate storage particularly to address enhanced resilience, emergency response, and renewable integration • Collaborate with industry on development of operations and control tools and algorithms that facilitate optimal utilization of storage • Researching mathematical models and algorithms for real-time optimal AC power-flow control and grid topology control optimization, including consideration of storage • Collaboratively address environmental uncertainties through partnered projects with the Dept. of Interior and the Army Corps of Engineers to improve water quality modeling and analysis tools for greater operational flexibility of pumped storage and hydropower projects
	• Collaborate with industry in prototype testing in controlled test bed(s) • Report results from ARRA projects incorporating storage. Collaborate with industry, States, DOD and other stakeholders on field trials and demonstrations of new or improved storage technologies, alternative deployment environments, enable evaluation of a range of grid applications/services or explore grid integration and operation/control approaches • Interface with private-sector financial institutions in the underwriting of innovative commercial energy storage projects applying to the DOE loan program.

Each Office will utilize their existing processes for engaging participants. These processes include workshops, funding opportunity announcements, small business innovation research grants, co-funding arrangements, and tasking within the Department or with National Laboratories. The coordination of these activities will take place on two levels: first, at the staff level where experts from individual Offices support and review activities undertaken by other Offices; and second, at the Assistant Secretary level, where coordination meetings and communications will ensure that all activities are focused on the major strategic objectives.

DOE supports research and development of a wide array of battery technologies (and other storage types). These technologies are at various levels of technical maturity and developmental risk. For those technologies that are relatively mature, the performance and cost attributes (as well as others like safety) are relatively well characterized, as are key areas where advancement will have the greatest potential impact on their deployment. The Energy Storage Handbook, cited on page 33, provides information to construct comparative mappings to other technologies and grid applications.

Figure 9 below shows, nominally, how several activities by different offices could work towards a solution, in this case for competitive storage technology.

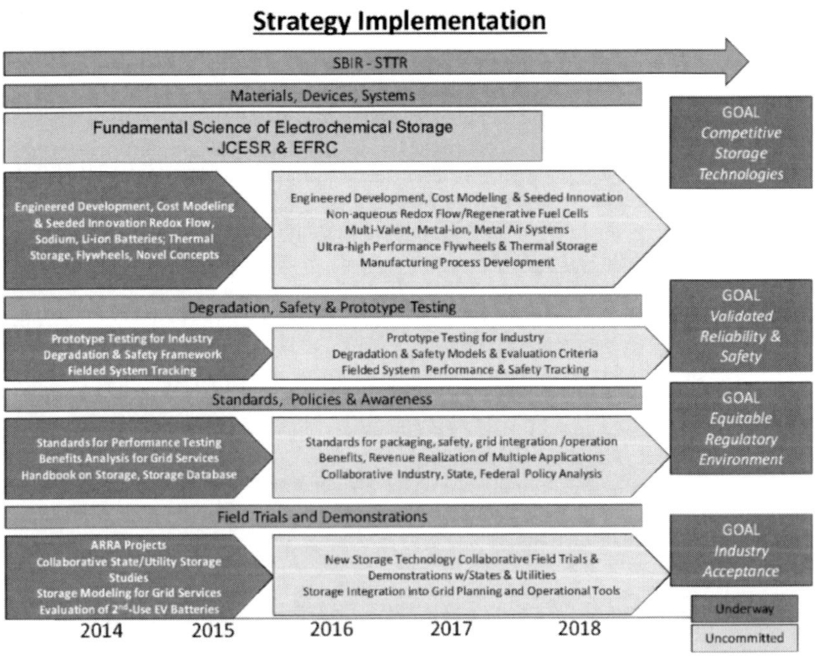

Figure 9. Summary Timeline of DOE Initiatives.

In the longer term:

- DOE can continue advancing material science and engineering, electrochemical sciences and engineering, systems engineering, and contribute to further cost reductions and performance improvement of an array of storage technologies. It is expected that the battery portion of this work can focus on such research that will shift toward technologies, like metal-air systems, that are currently at low technology readiness, as other storage technologies currently being developed are either transferred to the commercial sector or are proven to be unable to achieve competitive cost/performance. Power electronics efforts can expand into ultra-wide band gap materials to push efficiency bounds. DOE is also continuing to advance thermal energy storage and advanced heat transfer fluids for CSP applications.
- DOE can continue to refine our understanding of degradation and failure mechanisms, assist industry in translating that understanding into safer, more reliable and higher performing storage technology, and monitor fielded storage systems to factor experience into the DOE research and development efforts.
- DOE can resolve emerging standards challenges associated with accommodation of life- cycle testing, adaptation of standards and regulatory practices to new types of storage technologies, and harmonization of U.S. and international standards. DOE can monitor the evolving regulatory landscape to ensure its technology development and analytic efforts are consistent with and supportive of equitable regulatory practices.
- DOE can continue to collaborate with industry and other Federal and State entities to facilitate field testing of new storage technologies, update industry planning and operational tools to incorporate improved knowledge on grid storage technologies, and provide updated and improved models of the grid to enable optimization of storage system design.

Internal Collaboration

The DOE offices collaborate internally to raise the visibility of the issues, focus resources where they are needed and ensure that R&D results and industry needs are broadly communicated and employed to guide related R&D

among all offices. In addition to the already considerable collaboration between staff—e.g., where DOE experts from one office are asked to help evaluate and consider program direction in another office—there are formal collaboration forums, such as the Grid (Modernization) Technology Team, where programs are coordinated and sometimes jointly funded. Collaborative workshops that enable engagement of the R&D community, industry, States and regulators are held periodically to gain common appreciation of the energy storage development and deployment challenges and opportunities.

JOINT CENTER FOR ENERGY STORAGE RESEARCH (JCESR)

The JCESR is an innovation hub of the Department of Energy and led by Argonne National Laboratory. The JCESR brings together researchers from four additional DOE National Labs, five research universities, and four private firms. The Center advances next- generation battery technology through basic research using nanoscience tools, pursuing understanding of electrochemical materials and phenomena at the atomic and molecular level.

JCESR has focused research on several goals set for the next five years. By pursuing higher charge densities, new electrodes, and liquid suspension batteries, the group has a target of a five-fold increase in energy stored in today's batteries at one-fifth the cost. Currently, their research is focused on working with high-charge ions (magnesium & yttrium), improved chemical transformations, and improving non- aqueous redox flow (flow batteries). The science from JCESR will impact both transportation and grid applications.

While working together, as noted above, each office has clearly defined roles in the development of energy storage components and systems; these roles have evolved over time in alignment with the offices' core competencies. The Office of Science/Basic Energy Sciences (SC-BES), for example, conducts fundamental research into the scientific principles and physical processes underlying the material science and advanced electrochemistry necessary for storage technologies of the future in their "core" research projects, the six Energy Frontier Research Centers that focus on energy storage technology, and the Joint Center for Energy Storage Research—see box above.

The Advanced Research Projects Agency for Energy (ARPA-E) seeks high-risk, high-payoff projects which offer the possibility of significant and rapid developments in relevant technologies, including energy storage.

The Office of Energy Efficiency and Renewable Energy (EERE) focuses on the analysis of storage technologies as they apply to high penetration of renewable generation (e.g., wind, solar), addressing siting, permitting and environmental barriers to pumped hydro storage deployment, development of models to accurately characterize the capabilities of variable speed pumping technologies and the services they can provide, evaluate the techno-economic opportunities for the development of modular pumped hydro storage, and development of energy efficiency technologies particularly for commercial and residential buildings and transportation (including electric vehicles).

The Office of Electricity Delivery and Energy Reliability (OE) focuses on large-scale energy storage systems that can enhance the overall flexibility, reliability, resilience, and capability of the grid, and can enable transformation of the national electric generation and delivery system to meet the reliability and low-carbon emissions goals of the 21st century.

OE and ARPA-E Working Together on Energy Storage

Over the past decade, OE has led DOE grid-scale energy storage, through support of applied research, development, and demonstrations in partnership with companies, universities and national laboratories. ARPA-E was formed over the past 4 years with a focus on early-stage advanced research with high potential for impact across energy sectors. OE and ARPA-E work together to maximize impact and ensure development of new technology into energy storage applications. OE and ARPA-E collaborate on combined workshops; participation in interdepartmental working groups; co-participation in annual Peer-review; and combined on-site reviews for specific projects by OE and ARPA-E technology managers. Specific examples of technologies with coordinated support by ARPA-E and OE include:

Flywheel Energy Storage: OE supported research, development and deployment of flywheel energy storage technology, most notably for a 25kWh/15-minute storage unit.

A highlight of this effort is a pioneering ARRA-OE funded 20MW flywheel storage system for grid frequency regulation on the grid. in an array of 25kWh units. To provide a pathway to larger scale, lower cost flywheels at the 100kWh scale, ARPA-E has supported advanced research on higher energy density composite materials, flywheel magnetic levitation and advanced rotor dynamics. This work is in conjunction with an OE project to develop a 4x higher power flywheel drive will enable subsequent development of a 100kWh flywheel

Planar Sodium Metal Halide Battery: Sodium-beta alumina (Na-beta) battery is a chemical storage technology with the potential to ultimately meet cost and performance targets for renewable integration and grid applications. ARPA-E supported a national lab - company partnership to investigate a planar geometry Na- Beta battery for grid-scale energy storage. As a continuation, with support from OE, the national lab partner is currently working on understanding and mitigating the battery degradation mechanisms, which will ultimately result in increased safety, longer cycle life, and reduced cost.

Zinc Halide Flow Battery: Zinc halide based rechargeable liquid flow batteries could produce substantially more energy at lower cost than conventional batteries for grid storage applications. Under OE-ARRA support, this technology is being incorporated into a pioneering 25 MW / 3hr facility to firm 50MW of wind for a California utility. If successful, this storage technology will be demonstrated as a cost-effective alternative to 50MW of new generating capacity. ARPA-E supported laboratory scale research on an advanced electrode with mixed-metal catalyst materials, as an alternative to traditional carbon electrodes, providing greater durability and decrease cost.

Wide-Bandgap Semiconductor Power Electronics: OE supports development and demonstration of wide bandgap (WBG) semiconductor devices based on materials such as Silicon Carbide (SiC) and Gallium Nitride (GaN). Modules specifically are targeted for grid-tied energy storage systems which result in high power density and better performance than silicon-based systems.

ARPA-E supports development of WBG devices of increasing voltage, efficiency, integration and power handling capability, for electric power handling and management capabilities across a range of energy applications. This collaboration between OE and ARPA-E will enable more efficient storage management and control modules.

Through these related and integrated efforts on scientific research, technology development, demonstration and deployment, DOE is pursuing a coordinated agenda aimed at positioning the U.S. at the forefront of development of commercially relevant energy storage technology.

According to a recent GAO Report,[42] "DOE has taken steps to internally coordinate its battery and energy storage initiatives through activities that, among other things, defined common technology goals" (p.1) and "DOE has taken steps to internally coordinate its electric vehicle battery and electric grid storage initiatives through several actions. We found that these involved defining common technology goals; establishing strategies; and monitoring, evaluating, and reporting results. Specific steps… Established two working groups called integrated technical teams – Electric Vehicle Battery Technical Team and Grid Modernization Technical Team." (p.27)

External Collaboration

Importantly, DOE does not work in isolation from industrial organizations, other Federal and State agencies, and external stakeholders. Each Office maintains frequent and formal interactions with industry, academia, governments, regulatory bodies, and associations to ensure that priorities are re-calibrated and that resources are focused on the most urgent and impactful areas. This interaction with industry enables DOE to leverage its resources and expertise in advancing its overall mission to create a sustainable and commercially sound base of storage manufacturers and users.

Building and maintaining effective public–private partnerships is one of the key goals for achieving the objectives of the DOE's energy storage program. The strategy is to engage world- class professionals from key public and private organizations and help support and leverage research and development so that it meets the goals of DOE and the nation. Partners include electric utilities and manufacturers of energy storage devices, electricity consumers, project developers, and State and regional agencies.

Examples of electric utility stakeholders include investor-owned and public utilities; electric cooperatives; and Federally chartered entities such as the Tennessee Valley Authority, Bonneville Power Administration, and Western Area Power Administration. Partners also include the California Energy Commission and New York State Energy Research and Development Authority, who are partnering with major pioneering storage installations. DOE works closely with industry partners, and many of its projects (and all of its Recovery Act demonstrations projects) are cost-shared at a significant level. In the area of storage, DOE also works with national and regional interest groups, such as the Electricity Storage Association, National Rural Electric Cooperative Association (NRECA), National Alliance for Advanced Technology Batteries (NAATBaat), California Energy Storage Association (CESA), Texas Energy Storage Association (TESA), New York Battery Energy Storage Technology consortium (NYBEST), the storage association in Vermont, and the Pacific Northwest Economic Region – Energy Storage Coalition. DOE also has extensive collaboration with the National Electrical Manufacturers Association, and similar industry groups, in order to harmonize grid storage technology development and commercialization with appropriate industry standards and practices, both now and in the future.

The engagement of public–private partnerships takes several forms:

- Technical exchanges achieved through periodic conferences, workshops, annual peer reviews, informal meetings, and joint R&D planning sessions in addition to the work being executed by the various National Laboratories.
- Communications and outreach through websites, webcasts, meetings, and publications in technical journals to foster information sharing and technology transfer.
- Cost-shared R&D projects that leverage resources and focus on accomplishing tasks of mutual interest. It also signals the willingness of the private parties in taking over and sponsoring the energy storage efforts beyond the limitations of government.
- Competitive solicitations to engage the nation's top R&D performers in projects to design, fabricate, laboratory test, field test, and demonstrate new technologies, tools, and techniques.

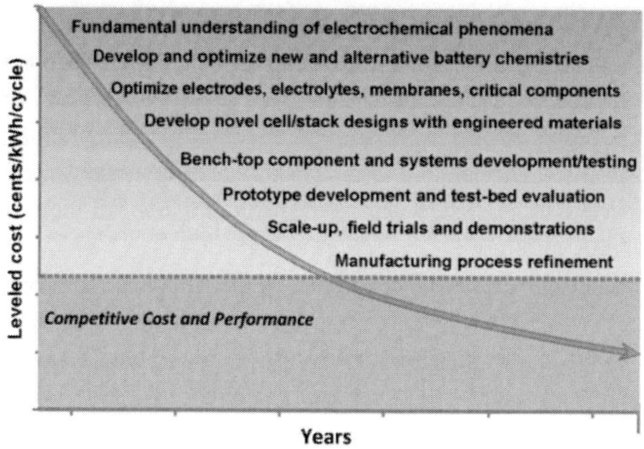

Figure 10. Steps to Drive Down Cost in Technology Development.

- Small Business Innovative Research (SBIR) grants, which can be used by Federal agencies to nurture innovative concepts from small businesses.
- International collaborations- DOE participates in the International Energy Agency Annex on Energy Conservation through Energy Storage, and collaborative projects that complement DOE efforts (e.g., Korea – sodium sulfur battery development).
- Inter-agency collaborations—DOE works with resource and regulatory agencies under the Federal Inland Hydropower Working Group to provide input to the creation and piloting a two-year FERC licensing process for closed-loop pumped-storage projects as called for in Section 6 of the Hydropower Regulatory Efficiency Act of 2013.

7.0. Actions Specific to Technology Development

As noted earlier, technology development spans a range of activity, depending on the nature of the technology, its maturity (e.g., technology readiness level), and anticipated application environment. As pointed out by stakeholders in multiple workshops, the process of accelerating development, maturation and deployment of storage technologies requires research, development and demonstration tailored to the technologies. An illustrative display of technology development (nominally for a battery) is shown below.

At all stages of technology development, improved scientific and engineering knowledge is vital for resolving critical cost and performance challenges as well as for feeding the creative processes for invention and innovation. Focused exploration of fundamental materials science, transport processes, and interface behavior relevant to energy storage contributes to the body of knowledge necessary for disruptive and evolutionary advances in energy storage technology.

In the strategy described above, there is an emphasis on advancements in research that can ensure long-term breakthroughs in cost and reliability, leading to a sufficient pipeline of future deployments. ***Applied material research***, enables the transition of scientific and engineering advances to specific challenges of critical components in energy storage devices. For the generalized battery development sequence depicted above, materials are modified with an understanding of electrochemical science and engineering fundamentals, e.g. current distribution, ion and electron transport, interfacial electron-transfer reactions, to enable greater energy density, higher current fluxes, slower dendrite growth, more conductive electrolytes, more effective separators, etc., enabling not only dramatic reductions in cost but also substantial improvements in device performance. Such advances are possible with many storage technologies, and development efforts are prioritized to components and concepts that will have the greatest impact on storage competitiveness, safety, lifetime, and other attributes, guided by detailed cost and performance models. One area of particular relevance is gaining a better understanding of degradation processes and mechanisms that affect the reliability of energy storage components. The ultimate goal is to effectively channel scientific advances and promising technologies towards commercialization.

In the area of ***device development***, the focus is on all the system components that comprise the AC energy storage system. The overall cost of energy storage systems can be roughly broken down to 25%-40% at the device level—i.e., where energy is held until needed, such as chemical/electrolytes in batteries, flywheels, etc.; and approximately 20% to 25% for associated power electronics—which ensure the proper and safe charge and discharge of stored energy; the remaining costs are associated with the balance of plant—i.e., facilities—as well as installation, engineering, and financing. Thus, R&D and demonstrations need to be undertaken in the development of design tools and electrochemical engineering for component and system tradeoff and optimization, as well as the development of new designs for components, subsystems, and systems. In this area we need targeted programs to reduce capital costs, at the device level.

Perhaps the greatest impact will come from ***system-level deployment***, since in situ testing and assessments can provide sufficient risk mitigation to further the adoption of storage by industry. DOE has undertaken a number of collaborative demonstrations with State and private partners. System-level deployments also include:

- Community and commercial establishments, university campuses, shopping centers, etc.
- Central station and distribution level
- Civilian and military microgrids
- Managing variability of renewable generation
- Isolated grids (Hawaii, Alaska, Texas, and islands)
- Grid services from EV/PHEV

The utility industry, as represented by the Electric Power Research Institute (EPRI)[43], has outlined use cases for demonstrations in the near future. Specifically the use cases involve: bulk energy storage systems (for renewable energy ramp control, resource adequacy, short-term balancing and reserves, etc.); distributed energy storage systems (for voltage regulation/reactive power support, peak load management, etc.); edge of grid energy storage systems (renewable integration, upgrade deferral, EV charging, back-up power); customer premise energy storage systems (electricity bill management, back-up reliability, renewable integration, upgrade deferral). A high priority is to have a demonstration focused on energy storage systems that can be deployed at electrical substations, in the 1-4 MW and the 3-4 hr size range. It will address the challenge of cost-effective deployment of such systems, providing multiple functions on the grid and provide multiple value streams to their owner/operators.

Demonstrations are vital contributors to both improving technology, but in tailoring to meet electric system requirements, and creating a body of experience that accelerates early adoption and supports resolution of institutional issues affecting deployment. DOE provided ARRA funding for extensive deployment activities. However, new and improved storage technologies are moving through the development pipeline and it may be necessary for these technologies to undergo field trials and demonstrations in order to facilitate their advancement to equitable consideration for deployment. Field demonstrations can be leveraged to validate application performance and model results to achieve the energy storage objectives.

DOD power systems for bases, in particular in microgrid formats, serve as a particularly valuable test and demonstration environment for new technology.

Associated with system-level deployment is a focus on working with industry, at the pre- competitive stage, to promote advancements in manufacturing and use cases. As we think further about the widespread deployment of energy storage systems, we need to focus on the manufacturability of both devices and system components, since advancements in manufacturing will allow for cost reductions and greater reliability, as well as addressing the need for smarter sourcing of resources and recycling existing materials.

Emergency Preparedness

A more reliable, resilient, and secure power system is essential for the protection of critical infrastructure across regions. Microgrid systems combined with grid scale energy storage are being developed as a robust solution for increasing the resiliency of critical infrastructure. Grid scale energy storage, when combined with distributed renewable generation, would allow microgrids to provide reliable power for essential services over an extended time period of emergency. During non-emergency time periods the system can reduce demand charges for the user and provide compensated services to the grid.

DOE has already initiated work with the States and with DOD to implement a number of such resilient microgrid designs. To further develop this storage/microgrid concept, DOE is collaborating with State energy offices, regulators and the private sector to develop and promote grid scale energy storage with microgrids enhancing resiliency of critical infrastructure.

8.0. Goals and Actions Specific to Analysis

Quantitative analytics is a critical component of an integrated storage RD&D strategy. It provides valuable cost/performance targets of storage systems and components for sustained market competitiveness, and insights into the market design barriers and regulatory impediments. It informs and supports decision makers at DOE and other funding agencies in guiding the RD&D agenda toward setting near-term and long-term goals for market introduction and prudent growth potential in a changing electricity market place as other new clean technologies are competing for market adoption. Analytical methods are applied to both forward-looking explorations and studies that consider future grid operations, market rules and environmental constraints, as well as on demonstrations and evaluations of technology prototypes in many different real-world applications and locations to substantiate and validate cost-performance characteristics of today's technologies.

Analytics is also being utilized to develop new or enhance existing design and engineering tools to consider the specific characteristics of storage technologies. Analytics methods enhance the generation resource and transmission planning process such that energy storage can be employed on an equal level with generation expansion, transmission and distribution system upgrades rather than as an afterthought once new generation and transmission has been deployed. This requires the integration of analytics methods into existing planning, design, and sizing tools as well as the education of today's and tomorrow's engineers in using these advanced tools.

Furthermore, market conditioning activities are necessary for the market introduction of advanced storage technologies. Activities focus on the development of codes and standards to measure performance, assure the health and safety in the deployment of novel storage systems, and provide communications and control protocols for the integration into existing energy management control networks.

Specifically, quantitative analytics enables decision makers and the nascent storage industry to:

- Sharpen cost and performance targets for various stationary storage applications and different market designs, locations within the electricity infrastructures, and use-cases
- Develop and enhance storage component cost modeling to guide the R&D agenda and reveal cost-reduction pathways for the vendor community.
- Articulate the value proposition(s), and develop business cases to instill confidence in nascent technologies among investors, regulators, and utility decision makers
- Assess the impact of grid operational differences across the regions, as well as different market designs on energy storage size and controls

requirements, and evaluate the role of storage in achieving a more resilient, reliable, cleaner energy infrastructure to increase the nation's energy security.

- Inform states and federal regulators and policy makers regarding the role of storage in grid operations, resource and transmission planning to meet the nation's needs for the 21st century.
- Develop or enhance planning, design and engineering tools to integrate storage into buildings, distribution, and transmission networks.
- Develop codes and standards rapidly to fill the gap of uniform procedures and guidelines necessary to wide-scale deployment and market acceptance.

To meet these outcomes above, the following targeted analyses and tool development activities are planned:

Component-Cost Modeling

DOE has already funded the initial development of a component cost model for redox flow batteries, fuel cells, Li-ion batteries, and other technologies to reveal the cost-reduction potential of individual components and the entire storage system and pathways to achieve cost/performance targets. Component-cost modeling would continue to capture cost improvement mechanisms by advancements in manufacturing, novel materials, and engineering designs. Component cost tools would be open-sourced and continually updated with new materials, engineering, and manufacturing approaches and for new electro-chemistries. This activity would benefit the vendor community, as well as inform DOE's program managers in adjusting the future RD&D agenda.

Electric Power System Analysis and Technical Support

To value the key system benefits of very fast and accurate response of energy storage technologies to grid operators' dispatch signals, a set of system analysis studies are envisioned that would specify new regulation reserve capacities that are science-based rather than rely on rule-of-thumb-based approaches. Additional analysis in collaborations with ISOs, and other grid operators would enable value-based treatment of reserve-assets.

Collaboration with States and Regions

The recent California PUC ruling on storage targets for 2020 represents a unique opportunity for increasing market adoption of electricity storage in California specifically, and the United States in general. Analytical work targeted to support the large-scale deployment of storage in all phases of the deployment including siting, selection of applications, operations and maintenance, would increase the likelihood of storage technology to become an enduring grid asset with sustainable business models.

More broadly, collaboration with FERC, RTO, state agencies, grid operators, and utility entities provides technical support to identify the specific regional grid operational challenges based on existing environmental constraints and future contributions from renewable energy resources.

Analysis of Demonstration Projects

Analytical studies would evaluate cost-effectiveness of high-value demonstrations to validate device performance and overall grid benefits. This activity would utilize the lessons-learned from the ARRA storage demonstrations and selected other applications of high value for the industry to solidify the expected performance and to create confidence in the investment and utility and grid operator communities. It is expected that this activity would be enduring along- side technology R&D program elements under various DOE funding mechanisms. This analytics element lends itself to cost-sharing with local and regional stakeholders.

Planning Tools Development

Today's engineering planning and design tools lack the ability to size, design, and validate the proper functioning of distribution system and transmission system operations with energy storage included in the mix of grid assets. It is unlikely that this gap in the tool-sets of grid planners would be filled by industry, particularly, not in time for the early market adoption phase in the next 3-5 years. Therefore, it is crucial for the DOE to provide the necessary analytical resources to develop new or enhance existing tools in collaboration with software developers and the planning communities. In particular, there is a need in both distribution and transmission planning tools to explore scenarios that maximize the utility of grid assets for multiple services. This requires complex co-optimizations of several objectives to be considered. If successfully implemented services and locations can be identified that allow utility and grid planners to find optimal locations and optimal control strategies that maximize the total value of storage.

Market Acceptance

This activity consists of codes and standards development engagements and outreach activities to regulators and market designers. This activity is envisioned to be an enduring component in the entire DOE near-term and long-term strategy. New technology development gaps in the existing codes and standards landscape across national and international standards may need to be evaluated to identify potential impediments of existing rules that would provide market entry barriers for novel designs and applications of energy storage.

9.0. Energy Storage Technology Standardization

As the energy storage field is developing a level of maturity and performance to the storage community and users, standardization is becoming paramount. A number of ongoing DOE activities already support this:

- The **DOE International Energy Storage Database, (IESDB)**[44] is the first freely accessible database of energy storage installations and related state and federal legislation/policies. It is an online tool designed to be accessible to a wide variety of stakeholders and has tremendous potential to help grow the energy storage industry. The IESDB is quickly becoming the go- to source for energy storage project and policy information.
- The **DOE/EPRI Electricity Storage Handbook in Collaboration with NRECA**[45] is a how- to guide for utility and rural cooperative engineers, planners, and decision-makers to plan and implement

electricity storage projects. Additionally, the handbook is an information resource for investors and venture capitalists on the latest developments in technologies and tools to guide their evaluations of electricity storage opportunities. It builds upon the *EPRI- DOE Handbook of Energy Storage for Transmission and Distribution Applications,* released in December 2003. It includes a comprehensive database of the cost of electricity storage systems in a wide variety of popular electric utility and customer applications, along with interconnection and integration schematics. A list of significant past and present electricity storage projects is also included for a practical perspective.

- **SNL Energy Storage Test Pad** provides third party cell-to-module-to-system validation of commercial and research scale storage solutions with a focus on grid scale applications. The lab and its staff provide engineering consultation on determining the most appropriate testing and applications for a given technology, performance and safety evaluation of the technology and systems, and provide public reporting of results for use by vendors, potential investors and customers.
- **Development of a protocol to measure and report performance of energy storage technology,** SNL is actively developing and distributing protocols for the energy storage community to provide uniform best practices for measuring, quantifying, and reporting the performance, reliability, applications and safety of energy storage systems.

APPENDIX A: SHORT DESCRIPTION OF DOE OFFICE PROGRAMS ON STORAGE

Office of Science

Office of Science, Basic Energy Sciences

The key research areas to realize advanced storage technologies for the future were identified in the SC-BES workshop on Basic Research Needs for Electrical Energy Storage (April 2007). DOE technology offices that participated in this workshop include Electricity Delivery and Energy Reliability (OE) for utility-scale energy storage and Energy Efficiency and

Renewable Energy (EERE) for their Vehicle Technologies program. A number of specific areas of fundamental research for both batteries and electrochemical capacitors are outlined in the associated workshop report *Basic Research Needs for Electrical Energy Storage* (http://science.energy.gov/~/media/bes/pdf/reports/files/ees_rpt.pdf). These include: (1) Novel designs and strategies for chemical energy storage, (2) Solid-electrolyte interfaces and interphases, (3) Capacitive energy storage materials by design, (4) Electrolyte interactions in capacitive energy storage, (5) Multifunctional materials for pseudocapacitors and hybrid devices, and (6) Theory and modeling.

Core Research

SC-BES funds approximately $7.5M/yr in research projects from the "core" single investigator and small group research program that, while fundamental in nature, address aspects of these scientific issues and are topically connected to electrical energy storage. In addition, SC-BES has made major investments that support energy storage research in the Energy Frontier Research Center (EFRC) program and the Batteries and Energy Storage Hub.

Energy Frontier Research Centers

In FY2009, SC-BES significantly increased the investment in fundamental energy storage research with the initiation of six EFRCs directly related to electrochemical energy storage (these EFRCs have a five-year budget of $90M). EFRC research is addressing the priority research directions outlined in the workshop report and includes investigations of interfacial properties and chemistry, development of high-performance nanostructured materials, understanding the role of electrolyte chemistry, discovery of novel electrocatalysts, and enhancing ion and electron transport properties and dynamics.

In addition, research in a number of cross-cutting EFRCs, while not directly underpinning batteries or capacitors, is addressing fundamental science that would propel electrical energy storage advancements (synopses of all 46 EFRCs can be found at http://science.energy.gov/bes/efrc/). Ongoing coordination activities with the DOE technology offices maintain awareness across DOE of the directions and outcomes of these research programs.

Energy Innovation Hub – Batteries and Energy Storage

In FY 2013, SC-BES announced the selection of the Joint Center for Energy Storage Research (JCESR), led by Argonne National Laboratory, for the Batteries and Energy Storage Hub. As defined in the funding opportunity announcement, the Batteries and Energy Storage Hub addresses the scientific and engineering research needed to advance the next generation of electrochemical energy storage for both transportation and the grid. The JCESR team brings together leading experts in both transportation and grid storage research to focus on the scientific barriers for electrochemical energy storage for both of these applications. JCESR benefits from co-location of major research activities funded by EERE, OE, and ARPA-E, facilitating the evolution of synergies among the programs and ensuring continued delineation of the energy storage research activities across DOE.

JCESR has launched a research program to perform underpinning science with the goal of developing research prototypes for electrochemical energy storage systems beyond lithium-ion for the grid and transportation. The research emphasizes discovery of new energy storage chemistries through the development of an atomic-level understanding of reaction pathways and development of universal design rules for electrolyte function. The overarching goals driving the scientific and engineering research towards next generation energy storage technologies are summarized by JCESR as "5/5/5": five times the energy density of today's systems at one-fifth the cost within five years, the award period for the Hub.

Federal management of the Hub is led by SC-BES in close coordination with EERE and ARPA- E for the initial selection review and for ongoing assessments of the Hub R&D activities. More information on JCESR can be found at http://www.jcesr.org/ and more information on Energy Innovation Hubs in general can be found at http://energy.gov/science-innovation/innovation/hubs

Energy Efficiency and Renewable Energy (EERE)

Energy Efficiency and Renewable Energy – Investments in Energy Storage

The Energy Efficiency and Renewable Energy Office is accelerating the research, development, demonstration, and deployment of new and advanced energy storage technologies that can be used in buildings, electric vehicles, or to help address the added variability and uncertainty of variable renewable technologies. Currently most of the work is concentrated in Sustainable Transportation and Renewable Power generation such as pumped hydro storage.

The Sustainable Transportation programs focus on reducing the cost, volume, and weight of batteries, while simultaneously improving the batteries' performance (power, energy, and durability) and ability to tolerate abuse conditions. Reaching the Office's goals in these areas and commercializing advanced energy storage technologies might allow more people to purchase and use electric drive vehicles.

- Exploratory Battery Materials Research: Addresses fundamental issues of materials and electrotechnical interactions associated with lithium ion and beyond lithium batteries. The research attempts to develop new and promising materials, use advanced material models to predict the modes in which batteries fail, and employ scientific diagnostic tools and techniques to gain insight into why materials and systems fail.
- Applied Battery Research: Focuses on optimizing next generation, high-energy lithium ion electrochemistries that incorporate new battery materials. This activity emphasizes identifying, diagnosing, and mitigating issues that negatively impact the performance and life of cells using advanced materials.
- Advanced Battery Development, Systems Analysis, and Testing: Focuses on the development of robust battery cells and modules to significantly reduce battery cost, increase life, and improve performance. This research aims to ensure these systems meet specific goals for particular vehicle applications.

The Renewable Power programs focus on using energy storage to address the variability and uncertainty of solar and wind and provide greater power system flexibility. This includes energy storage devices developed in concert with renewable power sources such as concentrating solar power or hydropower technologies. The Renewable Power programs focus in the following areas.

- Research and Development: This research investigates the development of new materials and systems such as the enhancement of thermal and thermochemical energy storage systems for concentrating solar power plants with the goal of driving cost and performance toward the SunShot targets.
- Testing and Field Demonstrations: This includes projects that investigate the efficacy and value of energy storage in addressing

issues such as variability and load shifting using technologies such as electrochemical storage or advanced pumped storage hydro plants.

- Modeling and Analysis**:** These efforts investigate the cost and benefits associated with using energy storage to enable the use of more renewable energy technologies in the electricity system. These include system wide studies in areas like Hawaii in the Hawaii Solar Integration Study, the West in the Western Wind and Solar Integration Study, and the United States in the Renewable Energy Futures Study and the SunShot Vision Study. This also includes an analysis of the emissions and economic benefits of hydrogen generation from wind power plants for use in fuel cells. Additionally, these efforts work to improve modeling capabilities as they relate to new technologies such as variable speed pumped hydro storage.

Advanced Research Projects Agency – Energy (ARPAE)

GRIDS

ARPA-E's Grid-Scale Rampable Intermittent Dispatchable Storage (GRIDS) program focuses on development of low-cost storage technologies for the electric grid. Specifically, GRIDS technologies are designed to address the challenge of renewable ramping. Initiated in 2010, ARPA-E's GRIDS program is developing new storage technologies with a capital cost of less than $100 per kilowatt hour that can scale to store megawatt-hours of electricity and be used at any location on the grid. The program includes 12 projects in a variety of technologies. A few of these are in conjunction with OE ARRA projects.

HEATS

ARPA-E's High Energy Advanced Thermal Storage (HEATS) program seeks to develop revolutionary, cost-effective ways to store thermal energy. Thermal energy--or heat energy--is involved in over 90% of all energy technologies, and there is a critical need to find efficient, cost-effective ways to store it. Initiated in 2010, the HEATS program includes 6 projects.

SBIR/STTR 2012

In 2012, ARPA-E funded a Small Business Innovation Research (SBIR) and Small Business Technology Transfer (STTR) program that included four energy storage projects focused on developing low-cost energy storage devices for use on the customer side of the meter. Low-cost storage located “behind the meter” can convert intermittent distributed generation assets into reliable low-cost power, as well as helping to shave peak loads on the centralized grid. ARPA- E’s SBIR/STTR program funded development of systems with peak power of 2.5 kW and duration of 4 hours with a footprint equivalent to that of an appliance (such as a refrigerator) that could be deployed in homes and small businesses as distributed energy storage resources.

ADEPT

ARPA-E's ADEPT program, short for "Agile Delivery of Electrical Power Technology," includes 14 projects in advanced power conversion technologies: circuits, transistors, inductors, transformers, and capacitors. Around 30% of the electricity used in the U.S. today flows through power converters, and a significant amount of energy is lost when that electricity is modified to a different current or voltage. That's because most power converters are based on decades-old technology and rely on expensive, bulky, and failure-prone components. The DOE estimates that in 20 years, 80% of the electricity used in the U.S. will flow through power converters, so there is a critical need to improve their efficiency. Initiated in 2010, the ADEPT program funds aim to develop advanced power conversion technologies that, if successful, would increase energy efficiency, improve the performance of electrical devices, and accelerate growth of the smart grid. ADEPT program innovations also could help establish U.S. businesses as technical leaders in power electronics, and bring lower power bills and less expensive electronics to American consumers.

SWITCHES

ARPA-E is currently seeking proposals for its SWITCHES program, short for “Strategies for Wide-Bandgap, Inexpensive Transistors for Controlling High Efficiency Systems.” Initiated in 2013, this program seeks to fund transformational advances in wide bandgap (WBG) materials, device fabrication, and device architectures. The goal of the SWITCHES program is to enable the development of high voltage (1200V+), high current (100A) single die power semiconductor devices that, upon ultimately reaching scale, would have the potential to reach functional cost parity with silicon power transistors while also offering breakthrough relative circuit performance (low losses, high switching frequencies, and high temperature operation). If successfully developed, these transformational technologies could reduce the barriers to ubiquitous deployment of low-loss WBG power semiconductor devices in stationary and transportation energy applications. SWITCHES technologies can increase energy efficiency, improve the performance of electrical devices, and accelerate growth of the smart grid.

Other Grid Storage Projects

ARPA-E funded the development of a number of grid storage technologies, including novel batteries and flow batteries, as part of its open solicitations in 2009 and 2012. Projects include work on sodium beta batteries, iron flow batteries, potassium and sodium based systems.

Office of Electricity Delivery and Energy Reliability

Office of Electricity Delivery and Energy Reliability – Energy Storage Program

The Energy Storage Program is focused on accelerating the development, demonstration and deployment of new and advanced energy storage technologies that will enhance the stability, reliability, resilience and economics of the future electric grid, which includes substantial contribution of intermittent renewable energy resources such as wind and solar power generation. The Energy Storage Program addresses three key challenges that are currently limiting widespread adoption of energy storage technologies.

- **Improving the cost/benefit ratio** of energy storage through advancements in materials engineering and device architectures;
- **Field Validation** of first-of-a-kind systems in representative utility environments to optimize storage devices for diverse utility applications and gain experience with the performance and behavior of diverse storage technologies;
- **Modeling and Analysis of Storage Systems** to assess the use, costs and benefits of energy storage, identify institutional and policy barriers, and develop tools for utilities and users planning to introduce and use energy storage.

To address these challenges, OE supports work by Universities, National Laboratories and Industry focused on four areas including:

- **Storage System R&D** – Addressing critical technical and economic limitations of state of the art energy storage technologies, and exploring new technologies that promise major advances in cost, performance, cycle life, and safety. Technologies primarily include mechanical (e.g. flywheels) and electrochemical (advanced redox flow, sodium, lithium, lead-acid, and novel derivatives and variations) batteries. Targeted material science and engineering R&D is conducted in order to maximize the durability, reliability and performance of the storage system while minimizing the installed capital cost.

- **Power Conversion and Grid Integration** – Addressing the unique power conversion needs for effective delivery of power to and from energy storage systems to enable maximum utilization of energy storage for a myriad of grid applications. Technical efforts focus on new power electronic components and converter architectures that reduce cost, improve performance, and extend life of power converters for energy storage.
- **Testing and Field Demonstrations** – Collaborating with manufacturers, States, municipalities, and utilities, in cost-shared testing and field demonstrations to establish operational performance of energy storage systems in real world conditions. These efforts focus on understanding and analyzing the benefits of deploying and utilizing energy storage to serve a wide range of application from: frequency regulation, load following, power quality enhancement, voltage and VAR support, load shifting, grid congestion relief, arbitrage, and many others. Siting issues, control strategies, integration with grid operations, maintenance, storage performance, as well as beneficial grid impacts are just a few issues uniquely addressed by field demonstrations. OE has responsibility for reporting on technical performance of ARRA-funded storage demonstration projects.
- **Analytic Studies** - Developing analytic tools of storage system, device, and component control and performance, storage cost models, new industry standards, value stream determination, and studies of market economics, and deployment and grid integration challenges, serves to inform stakeholders and guide R&D investments.

SBIR

OE participates in SBIR on an annual basis with a storage topic jointly managed with SC-BES. Projects involve advanced work on electrodes, electrolytes, and membranes.

ARRA

OE manages the ARRA program of Energy Storage Demonstrations. The program includes 16 projects representing Large Energy Storage for Renewable Integration, Frequency Regulation, Distributed Storage for Grid Support, Compressed Air, and Demonstration of Promising Storage Technologies. Federal Funding $185M for these projects has attracted a cost-share of $585M. In addition, OE also manages 7 ARRA Smart Grid projects involving application of energy storage.

Other Federal Agencies – Energy Storage R&D Activities

While DOE provided the largest amount of funding for energy storage research, several other agencies also had initiatives that examined Batteries and energy storage topics. Table A1 shows the net funding obligations of other federal agencies:

The following describes the initiatives that are relevant to grid-connected storage deployment for each agency.

Department of Defense

DOD has been making large investments in storage research each year, devoting roughly $145 million per year since 2009, representing between 21% and 28% of the electricity research budget. There are currently programs being run by each branch of the military, as well as jointly with other organizations (e.g., ESTCP/SERDP). Currently, the DOD is running several microgrid scale energy storage projects involving battery technology coupled with renewable and locally fired natural gas plants. These tests each focus on different technology aspects, battery types and chemistries, and battery size (large batteries versus distributed assets. Projects are focused on improving microgrid self-sufficiency (islanding) by improving load management and generation efficiency through the addition of energy storage. Programs in various stages of completion exist at Forts Hickam and Devens (in collaboration with DOE), Sill, Oklahoma; Marine Corps Air Station Miramar, California; 29 Palms, California; Portsmouth Naval Shipyard, Maine; Fort Bragg, North Carolina; and Fort Detrick, Maryland

In addition to the microgrid-level programs, several ongoing energy storage projects that could impact grid-level technologies:

Table A1[46]: Agency Battery and Energy Storage Initiatives and Funding Obligations, Fiscal Years 2009 through 2012 Agency

Agency	Number of initiatives	Funding obligations
DOE	11	$851,994,808
DOD	14	$430,274,229
NASA	8	$20,811,374
NSF	4	$8,582,868
EPA	1	$3,258,029
NIST	1	$1,375,000
Total	**39**	**$1,316,296,308**

- The Installation Energy Test Bed at the Marine Corps Air Station in Miramar currently testing new energy technology including battery technology (see microgrid discussion)
- Los Angeles Air Force Base is optimizing tool and implementation hardware for the strategic charging of their fleet of plug-in electric vehicles.
- DOD also determined that their bases have potential for 7 GW of Solar Energy generation; fully harnessing this potential would require significant amounts of energy storage capabilities
- US Army Corp of Engineers Engineer Research & Development Center (ERDC) is part of the ESTCP program (associate with SERDP-ESTCP) that advances energy storage technologies
- Naval Surface Warfare Center Crane Division has a joint program with Purdue University that grants a master's degree in chemical engineering for work on energy storage, and is conducting significant research into energy storage
- There are several organizations under the Army's Research, Development, & Engineering Command (RDECOM) that have energy research ongoing:
    - The Tank, Automotive Research Development Engineering Center (TARDEC) has several programs researching battery life and performance, information that may be applicable for distributed deployment performance and optimization
    - US Army Armament Research, Development & Engineering Center (ARDEC) currently has a partnership with Sandia National Laboratory and the DOE Energy Storage Program; some

of the joint efforts most recent breakthroughs involves better power electronics and power switches (thyristors)

- US Army Communications-Electronics Research, Development & Engineering Center (CERDEC) has been involved in a microgrid effort to provide a secure electricity grid at the Fort Belvoir 300 Area Compound site. This effort is done under the umbrella of the Energy Working Group (EWG), a joint effort between the DOD and DOE
- The Aviation & Missile Research, Development, and Engineering Center (AMRDEC), the Army Research Laboratory (ARL), and CERDEC have a joint program for collecting and storing solar power to provide continuous energy to company-level force protection systems in theater; the device relies on six rechargeable batteries
- The US Army Natick Soldier Research, Development, & Engineering Center (NSRDEC) has worked on improving lightweight and portable energy generation and storage systems (mostly and solar/wind combined with batteries) that can be deployed by soldiers in forward operations
- The Army Research Laboratory (ARL) is carrying out both basic and applied research on energy storage, especially focused on nano composites for ultra-lightweight multifunctional structural-energy storage systems

National Science Foundation

The National Science Foundation (NSF) has several initiatives (e.g., FREEDM and CURENT Engineering Research Centers, the GOALI Program) in place to support the development of batteries and associated technologies. The following programs are active:

- Energy for Sustainability[47] – a program providing funding for fundamental research and education that will enable innovative processes for the sustainable production of electricity and transportation fuels. One of several interest areas includes advanced batteries for transportation which targets high-energy density and high-power density batteries suitable for transportation.

- Energy, Power and Adaptive Systems (EPAS)[48] - a funding program for projects focusing on electric power networks and grids, including generation, transmission and integration of renewable, sustainable and distributed energy systems; high power electronics and drives.
- Sustainable Energy Pathways[49] – a funding program for innovative and interdisciplinary basic research in science, engineering and education by teams of researchers for developing systems approaches to sustainable energy pathways. One of the topic areas is sustainable energy storage solutions.
- Renewable Energy Storage (RESTOR)[50] - Funding program through the Emerging Frontiers in Research and Innovation (EFRI) which awarded four grants on energy storage in 2010. Funded projects included a nanostructured capacitor, thermochemical routes to solar fuel production, novel compressed air storage for offshore wind energy storage and a regenerative hydrogen-bromine fuel cell system for energy storage.

REFERENCES

Akhil, A. A., Huff, G., Currier, A. B., Kaun, B. C., Rastler, D. M., Chen, S. B., … & Gauntlett, W. D. (2013). DOE/EPRI 2013 Electricity Storage Handbook in Collaboration with NRECA. *Sandia National Laboratories Report, SAND2013-5131*. Retrieved from: http://www.sandia.gov/ess/publications/SAND2013-5131.pdf.

Auer, J. & Keil, J. *State-of-the-art electricity storage systems: Indispensable elements of the energy revolution.* [Powerpoint slides]. Retrieved from: http://www.dbresearch.com/PROD/DBR_INTERNET_EN-PROD/PROD0000000000286166/State-of-the-%20art+electricity+storage+systems%3A+Indispensable+elements+of+the+energy+r%20evolution.pdf.

California Energy Commission. (2020). Strategic Analysis Of Energy Storage In California. Retrieved from: http://www.law.berkeley.edu/files/bccj/CEC-500-2011-047.pdf.

Carnes, Kelly H. *Walking the Walk: DOD's Role in Accelerating Renewable Energy Technology Development* [Powerpoint slides]. Retrieved from: http://e2s2.ndia.org/schedule/Documents/Abstracts/12669.pdf.

Denholm, P., Fernandez, S. J., Hall, D. G., Mai, T. & Tegen, S. (2012). "Energy Storage Technologies," Chapter 12. National Renewable Energy

Laboratory. *Renewable Electricity Futures Study*, Vol. 2, Golden, CO: National Renewable Energy Laboratory; 12-1 – 12-42.

DOE International Energy Storage Database (beta). (2013). Policies. Retrieved from: http://www.energystorageexchange.org/policies.

DOE International Energy Storage Database (beta). (2013). *Projects*. Retrieved from: http://www.energystorageexchange.org/projects.

Doetsch, Christian. *Future Energy Storage/Balancing Demand* [Powerpoint slides]. Retrieved from: http://www.oeko.de/files/aktuelles/application/pdf/20120913_panel4_doetsch.pdf.

---- "Energy Storage Market Outlook H1. (2013). A New Awakening." *Bloomberg New Energy Finance,* 28 June 2013, 1-31. *New Energy Finance.* Web. 28 Jul. 2013.

Eyer, J. & Corey, G. (2010). Energy storage for the electricity grid: Benefits and market potential assessment guide. *Sandia National Laboratories Report, SAND2010-0815*, Albuquerque, New Mexico.

Hauer, A. *Energy Conservation Through Energy Storage Program* [Powerpoint slides]. Retrieved from: http://www.iea-eces.org/files/090525_broschuere_eces.pdf.

IEA- Energy Conservation through Energy Storage. *Annual Report 2012*. Retrieved from: http://www.iea-eces.org/files/annualreport_2012_1.pdf.

Lott, M. C., Kim, S. -I., Tam, C. & Elzinga, D. (2013). Technology Roadmap: Energy Storage. *Forthcoming IEA Working Paper, draft July 19th 2013.*

Narayanan, S. R., Prakash, G. K., Manohar, A., Yang, B., Malkhandi, S. & Kindler, A. (2012). Materials challenges and technical approaches for realizing inexpensive and robust iron–air batteries for large-scale energy storage. *Solid State Ionics*, *216*, 105-109.

Nicholson, M. & Stepp, M. (2012). "Lean, Mean, and Clean II: Assessing DOD Investments in Clean Energy Innovation." *The Information Technology & Innovation Foundation*. Retrieved from: http://www2.itif.org/2012-lean-mean-clean-dod-energy.pdf.

Rastler, D. M. (2010). *Electricity energy storage technology options: a white paper primer on applications, costs and benefits*. Electric Power Research Institute.

Rastler, D., Schainker, R., Kaun, B., Steeley, W. & Kamath, H. (2013). Grid Energy Storage: Challenges and Research Needs. *Forthcoming EPRI White Paper, Revision 7.*

SERDP-ESTCP Press Release. *DoD Study Finds 7,000 Megawatts of Solar Energy Potential on DoD Installations in Mojave Desert.* January 13th, 2013. Retrieved from: http://www.serdp.org/News-and-Events/News-

Announcements/Program-News/DoD-study-finds-7-000-megawatts-of-solar-energy-potential-on-DoD-installations-in-Mojave-Desert.

Teller, O., Nicolai, J. P., Lafoz, M., Laing, D., Tamme, R., Pedersen, A. S., … & Clerens, P. Joint EASE/EERA recommendations for a European Energy Storage Technology Development Roadmap towards 2030. *European Association for Storage of Energy.* Retrieved from: http://www.ease-storage.eu/tl_files/ease- documents/Stakeholders/ES%20Roadmap%202030/EASE- EERA%20ES%20Tech%20Dev%20Roadmap%202030%20Final%202013.03.11.pdf.

Turner, Guy. *Global Renewable Energy Market Outlook 2013 Fact Pack* [Powerpoint slides]. Retrieved from: http://about.bnef.com/fact-packs/global-renewable-energy-market-outlook-2013-fact-pack/.

US Government Accountability Office. *Batteries And Energy Storage: Federal Initiatives Supported Similar Technologies And Goals But Had Key Differences.* August 30, 2012. Retrieved from: http://www.gao.gov/assets/650/647742.pdf.

Wesoff, Eric. (2013). *Primus Lands DOD Contract for Energy Storage at Marine Base Microgrid.* Retrieved from: http://www.GreentechMedia.com

Yang, Z., Zhang, J., Kintner-Meyer, M. C., Lu, X., Choi, D., Lemmon, J. P. & Liu, J. (2011). Electrochemical energy storage for green grid. *Chemical Reviews*, *111*(5), 3577-3613.

End Notes

[1] http://www.energystorageexchange.org/ (All data cited in this paragraph is current as of August 2013). Note that the database has only verified the details of 121 of these deployments, with the details on the remaining projects in various stages of verification.

[2] For a table of several such estimates, see Hostick, D.; Belzer, D.B.; Hadley, S.W.; Markel, T.; Marnay, C.; Kintner- Meyer, M. (2012). End-Use Electricity Demand. Vol. 3 of Renewable Electricity Futures Study. NREL/TP-6A20-52409-3. Golden, CO: National Renewable Energy Laboratory.

[3] The bills before congress are S. 1030 (STORAGE Act) and S. 795 (MLP Parity Act). Details on the California bill (AB 2514) can be found on the CPUC website: http://www.cpuc.ca.gov/PUC/energy/electric/storage.htm

[4] Fitch Ratings, "Frayed Wires: US Transmission System Shows Its Age," 2006

[5] "Economic Benefits Of Increasing Electric Grid Resilience To Weather Outages" August 12, 2013. Available at: http://energy.gov/sites/prod/files/2013/08/f2/Grid%20Resiliency%20Report_FINAL.pdf

[6] See also "Storm Reconstruction: Rebuild Smart: Reduce Outages, Save Lives, and Protect Property," NEMA, National Electrical Manufacturers Association, 2013; and "

Recommendations to Improve the Strength and Resilience of the Empire State's Infrastructure," NYS 2100 Commission, 2012.

[7] IMS Research (now owned by IHS-CERA) report 'The Role of Energy Storage in the PV Industry – World – 2013 Edition'.

[8] U.S. Department of Energy 'Strategic Plan May 2011' (http://energy.gov/sites/prod/files/2011_DOE_Strategic_Plan_.pdf) page 15 and page 17

[9] : http://www.energystorageexchange.org/ (All data cited in this paragraph is current as of August 2013)

[10] Note that the database has only verified the details of 121 of these deployments, with the details on the remaining projects in various stages of verification.

[11] This information also was accessed in August 2013, and can be found at: http://www.energystorageexchange.org/

[12] Policy information come from the Bloomberg New Energy Finance report on storage dated June, 2013 and the Sandia National Laboratories database: http://www.energystorageexchange.org/

[13] See Bloomberg New Energy Finance H1 2013, page 8.

[14] S. 795 and S. 1845

[15] Note that this number includes some of the projects funded by the 2009 ARRA that have yet to come online; these projects total 334MW, or roughly 1/4th of the total target.

[16] http://www.capitol.state.tx.us/billlookup/history.aspx?legsess=82r&bill=sb943

[17] http://www.ny-best.org/

[18] HB 1289 and HB1296

[19] http://www.nrel.gov/hydrogen/cfm/pdfs/arra_deployment_cdps_q12013_4web.pdf

[20] See Bloomberg New Energy Finance H1 2013, pages 19-22.

[21] See Bloomberg New Energy Finance H1 2013, page 23.

[22] http://www.saftbatteries.com/press/press-releases/saft-receives-%E2%82%AC35m-orders-reliance-jio-infocomm-limited-rjil-li-ion-telecom

[23] Information in this table comes from Bloomberg New Energy Finance's Energy Storage Market Outlook, June, 28, 2013, as well as the DOE database on Energy Storage Projects referenced earlier. Conversions based on 1 euro = $1.30

[24] Several technologies that are still in the early stages of research have been omitted, as they are unlikely to be commercially viable within the next 3-5 years.

[25] This is the Prudent Energy VRB-ESS® - Gills Onions, California. The information on US installations comes from the DOE Energy Storage database referenced earlier.

[26] Source: http://web.anl.gov/energy-storage-science/publications/EES_rpt.pdf

[27] A grid service, or application, is a use whereas a benefit connotes a value. A benefit is generally quantified in terms of a monetary or financial value.

[28] A more comprehensive discussion of energy storage applications at all levels can be found in two documents referenced elsewhere in this report: Eyer (2010) and Chapter 1 of the DOE/EPRI 2013 Handbook.

[29] "*Use of Frequency Response Metrics to Assess the Planning and Operating Requirements for Reliable Integration of Variable Renewable Generation*," Joseph H. Eto (Principal Investigator) et al., LBNL-4142E, Lawrence Berkeley National Laboratory, Berkeley, CA, December 2010. http://www.ferc.gov/industries/electric/indus-act/reliability/frequencyresponsemetrics-report.pdf

[30] Spinning reserve is defined in the NERC Glossary as "Unloaded generation that is synchronized and ready to serve additional demand."

[31] Non-spinning reserve is not uniformly the same in different reliability regions. It generally consists of generation resources that are offline, but could be brought online within 10 to 30 minutes and could also include loads that can be interrupted in that time window.

[32] Swings of more than 100 MW, roughly the same capacity as a small power plant, can occur within a single five- minute period. The peak 5-minute wind generation down-ramp

experience occurred in June 2008 by the Bonneville Power Administration, having a value of -725MW.

[33] Eyer and Corey, 14.

[34] Similarly, this strategy could facilitate taking equipment out of service for maintenance.

[35] ANSI C84.1 "American National Standard for Electric Power Systems and Equipment – Voltage Ratings (60 Hz)" establishes nominal voltage ratings for utilities to regulate the service delivery and operating tolerances at the point of use.

[36] Strategen reported that in California, the IRR for a 100kW 4 hr battery system, increased from 8.2%/yr to 18%/yr when storage was combined with PV generation, due to favorable incentive rate structures. "Energy Storage – Shaping the Future of California's Electric Power System", Prepared for DistribuTECH 2011, February 2, 2011, Strategen and California Energy Storage Alliance.

[37] For example, at a recent roundtable with several key stakeholders from utilities and energy companies, there was particular skepticism over claims of 20-years of high-efficiency charging/discharging that some battery manufacturers claim.

[38] 1) Electric Power Industry Needs for Grid-Scale Storage Applications, Prepared by Nexight Group, Sponsored by U.S. Department of Energy, Office of Electricity Delivery and Energy Reliability, and the Office of Energy Efficiency and Renewable Energy, Solar Technologies Program, and 2) Advanced Materials and Devices for Stationary Electrical Energy Storage Applications, Prepared by Nexight Group, Sponsored by U.S. Department of Energy Office of Electricity Delivery and Energy Reliability, and the Advanced Research Projects Agency, December 2010.

[39] For current cost information, see Chapter 2 of Akhil, A.A., Huff, G, Currier, A.B., Kaun, B.C, Rastler, D.M., Chen, S.B., … , Gauntlett, W.D. (2013). DOE/EPRI 2013 Electricity Storage Handbook in Collaboration with NRECA. *Sandia National Laboratories Report,SAND2013-5131.*

[40] For current cost information, see Chapter 2 of Akhil, A.A., Huff, G, Currier, A.B., Kaun, B.C, Rastler, D.M., Chen, S.B., … , Gauntlett, W.D. (2013). DOE/EPRI 2013 Electricity Storage Handbook in Collaboration with NRECA. *Sandia National Laboratories Report,SAND2013-5131.*

[41] National Assessment of Energy Storage for Grid Balancing and Arbitrage PNNL-21388 (2013)

[42] GAO, *Batteries and Energy Storage Federal Initiatives Supported Similar Technologies and Goals but Had Key Differences,* August 30, 2012.

[43] EPRI, *Grid Energy Storage: Challenges and Research Needs,* Draft White Paper, July 8, 2013.

[44] Available at: http://www.energystorageexchange.org/

[45] Available at: http://www.sandia.gov/ess/publications/SAND2013-5131.pdf

[46] Source: GAO-12-842, Batteries and Energy Storage

[47] National Science Foundation, "Energy for Sustainability," https://www.nsf.gov/funding/pgm_summ.jsp?pims_id=501026, accessed July 2013

[48] National Science Foundation, "Energy, Power, and Adaptive Systems (EPAS)," http://www.nsf.gov/funding/pgm_summ.jsp?pims_id=13380, accessed July 2013

[49] National Science Foundation, "Sustainable Energy Pathways (SEP)," http://www.nsf.gov/funding/pgm_summ.jsp?pims_id=504690, accessed July 2013

[50] National Science Foundation, "ENG/EFRI FY2010 Awards Announcement," http://www.nsf.gov/eng/efri/fy10awards_RESTOR.jsp, accessed July 2013

In: Grid Energy Storage
Editor: Jonatan Janko
ISBN: 978-1-63117-368-4

Chapter 2

ENERGY STORAGE TECHNOLOGIES*

P. Denholm, S. J. Fernandez, D. G. Hall, T. Mai and S. Tegen

12.1. INTRODUCTION

Energy storage is one of several potentially important enabling technologies supporting large- scale deployment of renewable energy, particularly variable renewables such as solar photovoltaics (PV) and wind. Although energy storage does not produce energy—in fact, it is a net consumer due to efficiency losses—it does potentially allow greater use of variable renewables by shifting energy from periods of low demand to periods of high demand, which reduces curtailment and eases integration challenges. Energy storage can also provide a variety of high value services such as firm capacity and multiple ancillary services.

Energy storage is used in electric grids in the United States and worldwide. It is dominated by pumped-storage hydropower (PSH), with about 20 GW[164] deployed in the United States and more than 127 GW deployed

* This is an edited, reformatted and augmented version of text that originally appeared as Chapter 12 in Renewable Electricity Futures Study, Vol. 2: Renewable Electricity Generation and Storage Technologies, published by National Renewable Energy Laboratory, June 2012. (Section numbers and footnote numbers have not been altered).

worldwide (EIA 2008; Ingram 2010). In the United States, PSH was built largely in response to market conditions in the 1970s, including high oil and natural gas prices, regulatory restrictions on plants burning oil and gas, dependence on low-efficiency steam plants for peaking power, and anticipated "build-out" of a largely inflexible nuclear fleet (Denholm et al. 2010). In addition to PSH, a single, 110-MW compressed air energy storage (CAES) facility has been constructed in the United States (EPRI/DOE 2003). CAES is described in Section 12.3.2.3.

Deployment of storage in the United States over the past two decades has been limited by low natural gas prices, availability of high-efficiency and flexible gas turbines, and limited cost reductions in storage technologies. In addition, the regulatory treatment of storage, costly licensing and permitting, challenges with storage valuation, as well as utility risk aversion (including market uncertainty) have also limited storage development (EAC 2008). Figure 12.1 shows the installations of bulk energy storage in the United States.

Interest in energy storage technologies, which has reemerged over the past decade, has been motivated by at least five factors:

- Advances in storage technologies
- Volatility of fossil fuel prices
- The development of deregulated energy markets, including markets for high-value ancillary services[165]
- Challenges to siting new transmission and distribution facilities
- The perceived need and opportunities for storage with variable renewable generators and their role to reduce carbon dioxide emissions.

Along with this interest, there have been a number of new proposals and demonstration projects. Table 12.1 lists several proposed or installed projects (since 2000). Although there is significant interest in batteries and CAES, PSH continues to be the dominant proposed storage technology.

12.2. Resource Availability Estimates

The ability to site certain storage technologies (conventional PSH and CAES) is based on specific geologic characteristics. These issues are discussed in the technology-specific sections (Section 12.3 and 12.4).

Table 12.1. U.S. Electricity Storage Facilities Installed or Proposed Since 2000

Technology	Primary Application	Size (MW)	Owner/ Developer	Location(s)	Status
PSH	Load leveling/ firm capacity/ ancillary services	>40,000	Various	Various (see Figure 12.9)	Proposed[a]
CAES	Load leveling/ firm capacity/ ancillary services	300	PG&E[c]	Kern County, California	Proposed
		150	NYSEG[d]	Reading, New York	Proposed
		2,700	FirstEnergy[e]	Norton, Ohio	Proposed
Sodium-sulfur (NaS) battery	T&D deferral/ congestion relief	1	AEP[f]	North Charleston, West Virginia	Installed (2006)
		2	AEP	Bluffton, Ohio Balls Gap, West Virginia East Busco, Indiana	Installed (2008)
		4	AEP	Presidio, Texas	Installed (2009)
		1	Xcel Energy[g]	Luverne, Minnesota	Installed (2009)
Vanadium redox battery	T&D deferral/ congestion relief	0.25	Pacificorp	Moab, Utah	Installed (2004)
Lithium-ion battery	Frequency regulation	1	AES/PJM Inter-connection	Valley Forge, Pennsylvania	Installed (2008)
Flywheel	Frequency regulation	20	Beacon[h]	Stephentown, New York	Installed (2011)
		1	Beacon	Groveport, Ohio	Installed (2008)
		1	Beacon	Tyngsboro, Massachusetts	Installed (2009)

[a] As of December 2011, FERC had issued preliminary permits for 4d plants, representing approximately 35 GW of capacity. The capacity of proposed plants (including those with issued and pending preliminary permits exceeds 40 GW) (FERC n.d.). A map of proposed locations is provided in Figure 12.9.

[c] H. LaFlash "Compressed Air Energy Storage" slide presentation, Pacific Gas and Electric Company, November 3, 2010, http://www.sandia.gov/ess/docs/pr_conferences/2010/laflash_pge.pdf

[d] J. Rettberg, "Seneca Advanced Compressed Air Energy Storage (CAES) 150MW Plant Using an Existing Salt Cavern," slide presentation, November 3, 2010, http://www.sandia.gov/ess/docs/pr_conferences/2010/rettberg_nyseg.pdf NYSEG.

[e] Norton Energy Storage (2000)

[f] Parfomak (2012)

[g] Xcel Energy, http://www.gridpoint.com/Libraries/Featured_Media_Coverage_PDFs/wind-to-battery_-_Xcel_Energy_Brochure.sflb.ashx

[h] Beacon Power Corporation, http://phx.corporate-ir.net/External.File?item=UGFyZW50SUQ9NDY1Mjd8Q2hpbGRJRD0tMXxUeXBlPTM=&t=1, http://phx.corporate-ir.net/External.File?item=UGFyZW50SUQ9MjAxNTh8Q2hpbGRJRD0tMXxUeXBlPTM=&t=1, http://phx.corporate-ir.net/External.File?item=UGFyZW50SUQ9ODI0OXxDaGlsZElEPS0xfFR5cGU9Mw==&t=1, http://phx.corporate-ir.net/External.File?item=UGFyZW50SUQ9MzczNDQxfENoaWxkSUQ9MzcxMjE1fFR5cGU9MQ==&t=1, http://www.beaconpower.com/company/news.asp

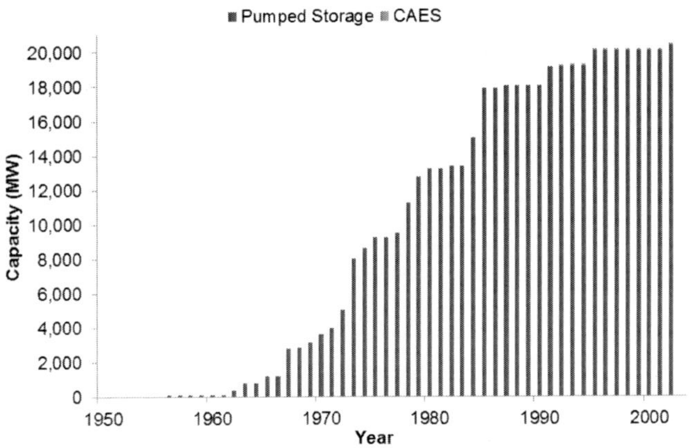

Source: EIA 2008.

Figure 12.1. Capacity of bulk energy storage systems in United States, 1956–2003.

12.3. Technology Characterization

12.3.1. Technology Overview and Applications

Energy storage technologies are typically characterized by their applications, often in terms of discharge time. Three common categories are provided in Table 12.2.

Table 12.2. Three Classes of Energy Storage

Common Name	Example Applications	Discharge Time Required
Power quality and regulation	Transient stability, reactive power, frequency regulation	Seconds to minutes
Bridging power	Contingency reserves, ramping	Minutes to ~1 hour
Energy management	Load leveling, firm capacity, T&D deferral	Hours

The first two categories of energy storage applications in Table 12.2 correspond to a range of ramping and ancillary services but do not typically require continuous discharge for extended periods. Storage technologies can provide local power quality benefits, such as voltage stability and provision of reactive power, and can increase the stability of the system as a whole by providing real or virtual inertia. As discussed in Chapter 4 (Volume 1), a high variable- generation grid will require increased operating reserves for frequency regulation due to short- term variability of the wind and solar resources; it will also require reserves covering forecast errors. Forecasting errors, especially over-prediction of wind or solar, requires time to allow fast-start thermal generators to come online. Hydropower and thermal units operating at part load typically provide operating reserves, but operating reserves can also be provided by energy storage technologies, often more efficiently or at a lower cost. Frequency regulation, for example, requires rapid response, and storage devices may provide faster response than traditional generators (Makarov et al. 2008). Storage technologies also have the unique ability to potentially provide reserves greater than their rated output while charging. A device charging at 1 MW can actually provide 2 MW of reserve capacity by stopping charging and rapidly switching to discharging; however, this ability is potentially limited by the technology-dependent switchover time. Previous analysis has demonstrated the potential benefits of providing fast ramping with energy storage to address the increase in sub-hourly variations resulting from large-scale deployment of variable generation (KEMA 2010).

The third category of services in Table 12.2 (energy management) corresponds to energy flexibility—the ability to shift bulk energy over several hours or more—which is the focus of storage deployment in the RE Futures scenarios.[166] An energy management device stores energy during periods of low demand (and correspondingly low energy prices) and discharges energy during periods of high demand and prices. In a high renewables scenario, this operation would be the same, and the charging and discharging periods would be driven by the combination of normal demand patterns and the supply of available variable generation. This includes storing energy when it might otherwise need to be curtailed due to low demand or constrained transmission. Storage devices sized for energy management can provide an alternative (or supplement) to developing new transmission capacity. Use of dedicated long-distance transmission for wind or solar power will be limited by the relatively low capacity factor of the resource. Storage could help reduce curtailment due to transmission constraints by co-locating storage with variable-generation sources and allowing them to increase use of transmission lines (Desai et al. 2003). This could also decrease the amount of new transmission needed, but represents a trade-off between the most cost-effective use of storage, and the cost of new transmission (Denholm and Sioshansi 2009). Figure 12.2 provides one example of the range of technologies available for these three classes of services and shows that many technologies can provide services across the timescales shown. Many energy management storage devices can provide fast response and provide power quality and bridging power services (the discharge times shown represent the continuous discharge capability as opposed to the response time).

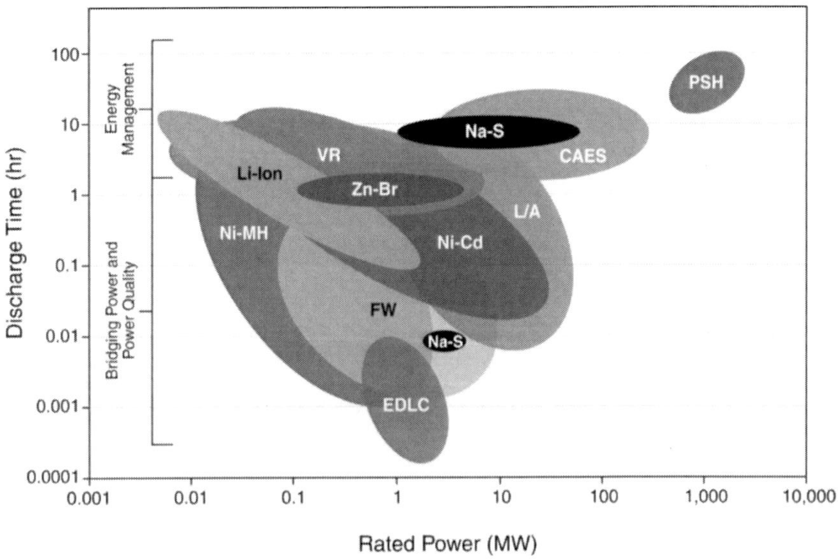

Source: Electricity Storage Association (ESA 2011)

System Ratings: Installed or proposed systems as of November 2008. This chart is meant to represent a general range of storage technologies and is not inclusive of all technologies, applications, and possible sizes.

CAES	Compressed air
EDLC	Dbl-layer capacitors
FW	Flywheels
L/A	Lead-acid
Li-Ion	Lithium-ion
Na-S	Sodium-sulfur
Ni-Cd	Nickel-cadmium
Ni-MH	Nickel-metal hydride
PSH	Pumped-storage hydropower
VR	Vanadium redox
Zn-Br	Zinc-bromine

Figure 12.2. Energy storage applications and technologies.

Figure 12.2 does not include thermal energy storage, which would cover a power range of a few kilowatts for thermal energy storage (TES) in buildings to more than 100 MW in concentrating solar power (CSP) plants, with a discharge time of minutes to several hours.

12.3.2. Technologies Included in RE Futures Scenario Analysis

Utility-scale electricity storage is modeled in the Regional Energy Deployment System (ReEDS) model to provide three services: firm capacity, energy supply shifting, and operating reserves. However, the primary grid integration challenge in a high renewable penetration scenario is the limited coincidence of renewables supply with normal electricity demand. Consequently, storage modeling for RE Futures focused on energy storage technologies that can provide energy management services or can store and discharge continuously for several hours (defined here as 8–15 hours, depending on the technology). This allows energy storage to u se otherwise potentially curtailed energy from variable-generation sources during periods of high generation and low load. As discussed later in this section, the modeling assumptions inherently undervalue shorter term and distributed storage devices, and they restrict their adoption; therefore, RE Futures cannot be used as an indicator of the opportunities for energy storage of all types.

Three technology groups meeting the criteria of being able to provide energy management services were included in the ReEDS modeling: high-energy batteries, pumped-storage hydropower, and compressed air energy storage. These technologies and their implementation in ReEDS are described in the following sections.

Notably absent from the modeling effort were short discharge and power quality applications such as flywheels and high power batteries. The most economic application for these devices appears to be fast-responding frequency regulation markets (Walawalkar et al. 2007). The ReEDS model combines frequency regulation and other reserves (for forecast error and contingency reserves), for example, into a single operating reserve constraint that can be provided by multiple technologies. Although RE Futures captures the increased need for operating reserves as greater levels of variable generation are deployed, it does not explicitly treat sub-hourly or sub-minute events (e.g., frequency regulation), and therefore cannot capture the high value of a regulation reserve device in isolation. As a result, although RE Futures can identify the overall need for reserves and the corresponding possible increase in the role of storage for operating reserves, it does not currently disaggregate the market and identify opportunities for individual reserve technologies. Recognizing this limitation, no attempt was made to estimate deployment of any individual reserve supplying storage technology.

In addition, because ReEDS is essentially a "bulk planning" model, it does not identify the potential value and opportunities of storage sited in the distribution system. In particular, it cannot evaluate opportunities to relieve local transmission or distribution congestion, or the value of T&D deferral. These applications are a primary application for current high-energy batteries such as flow batteries or NaS (Nourai 2007). This is also a primary application for end- use TES (ADM 2006). As a result, ReEDS will undervalue these and restrict their adoption into the marketplace.

Furthermore, the role of V2G was not explicitly evaluated in RE Futures. The RE Futures study included the value of controlled charging; however, uncertainty in the ultimate acceptance among original equipment manufacturers (OEMs), utilities, and consumers of V2G led to the conservative assumption to not include the potentially very large role of V2G.

Finally, limited deployment of hydrogen as a storage medium, and large uncertainty of cost- reduction and performance improvements of hydrogen storage, led to its exclusion as a core energy storage technology evaluated in RE Futures.

For these reasons, the ReEDS storage results are aggregated to show the total amount of storage deployed, as opposed to the deployment of individual storage technologies. RE Futures was used more to indicate the amount of bulk storage that may be beneficial to the grid (within the cost ranges and availability modeled) as opposed to evaluating particular storage technology types.

The particular energy storage technology deployed by ReEDS could actually be any of a number of storage technologies or an emerging technology not evaluated.

12.3.2.1. High-Energy Batteries

For many batteries, there is considerable overlap between energy management and shorter-term applications. Furthermore, batteries can generally provide rapid response, which means that batteries "designed" for energy management can potentially provide services over all applications and timescales discussed.

Several battery technologies have been demonstrated or deployed for energy management applications. The commercially available batteries targeted to energy management include two general types: high-temperature batteries and liquid electrolyte flow-batteries. Other commercially available battery types are generally targeted towards high-power applications and discussed in Section 12.3.4.

High-temperature batteries operate above 250°C and use molten materials to serve as the positive and negative elements of the battery. The most mature high-temperature battery as of 2011 is the sodium-sulfur battery (NaS), which has worldwide installations that exceed 270 MW (Rastler 2008). Several utilities have deployed the NaS battery in the United States.

Alternative high-temperature chemistries have been proposed and are in various stages of development and commercialization. One example is the sodium-nickel chloride battery (Baker 2008). The second class of high energy batteries is the liquid electrolyte "flow" battery. This battery uses a liquid electrolyte separated by a membrane (EPRI/DOE 2003). The advantage of this technology is that the power component and the energy component can be sized independently, with the electrolyte held in large storage tanks. As of 2011, there has been limited deployment of two types of flow batteries—vanadium redox and zinc-bromine. Other combinations such as polysulfide-bromine have been pursed, and new chemistries are under development (Yang et al. 2011).

In the United States, a primary focus of energy management batteries has been T&D deferral; however, demonstration projects have been deployed for multiple applications (Nourai 2007; EPRI/DOE 2003).

For RE Futures, batteries were combined into a single technology type, with performance based on a NaS battery; however, given the multiple battery types, and with uncertain cost reductions and technology improvements, the RE Futures battery technology should be considered a generic "high-energy" battery with 8 hours of discharge time. This could include technologies currently under various stages of development and deployment such as advanced lithium-based batteries. As with certain supply technologies, such as solar PV with multiple technology options, the goal was not to "pick winners" because the market will ultimately determine technology pathways based on cost and performance.

12.3.2.2. Pumped-Storage Hydropower

PSH is the only energy storage technology deployed on a gigawatt scale in the United States and worldwide. In the United States, about 20 GW is deployed at 39 sites, and installations range in capacity from less than 50 MW to 2,800 MW (EIA 2008). This capacity was largely built during the 1960s, 1970s, and 1980s (ASCE 1993). While there are a number of proposed plants, there has been no large-scale PSH development in the United States since 1995; however, development has continued in Europe and Asia (Deane et al. 2010). Lack of construction of new U.S. facilities has been largely been due to cost, market issues, and regulatory issues discussed in Section 12.1.

Pumped-storage hydropower stores energy by pumping water from a lower-level reservoir (e.g., a lake) to a higher-elevation reservoir using lower-cost, off-peak electric power. During periods of high electricity demand, the water is released to the lower reservoir to turn turbines to generate electricity, similar to the way in which conventional hydropower plants generate electricity.

Many existing PSH plants store 8 hours or more of energy, making them useful for load leveling, and providing firm capacity. PSH can also ramp rapidly while generating, making it useful for load following and providing ancillary services including contingency spinning reserves and frequency regulation (Phillips 2000).

Figure 12.3 shows a representative conceptual configuration of a PSH plant.

Pumped-storage hydropower plants often make use of an existing river or lake, avoiding the need for—and cost of—construction of a separate (usually the lower) reservoir. This is called an *open- cycle* PSH plant. In an instance in which a suitable natural water body is not available for use as one of the reservoirs, both the upper reservoir and the lower reservoir must be constructed. This type of construction is known as a *closed-cycle* plant, inasmuch as it has minimal interaction with natural water bodies. A water source is needed for a closed-cycle plant to provide water to initially fill the reservoir and compensate for losses during operation due to leakage and evaporation. Nearby rivers or streams are typical sources; treated municipal grey water or groundwater (wells) can also be used (Yang and Jackson 2011). Of the 45 PSH plants with preliminary permits from FERC, which include a total or more than 35 GW of capacity, at least nine have proposed closed-cycle PSH plants, and these exceed 9 GW of capacity (FERC n.d.).

12.3.2.3. Compressed Air Energy Storage

CAES stores energy by compressing air in an airtight underground storage cavern. To extract the stored energy, compressed air is drawn from the storage cavern, heated, and then expanded through a high-pressure turbine that captures some of the energy in the compressed air. The air is then mixed with fuel and combusted, and the exhaust is expanded through a low-pressure gas turbine. The turbines are connected to an electrical generator (Succar and Williams 2008).

CAES is based on conventional gas turbine technology and is considered a hybrid generation and storage system because it requires combustion in the gas turbine.[167] Instead of a round-trip efficiency number, the performance of a conventional CAES plant is based on its energy ratio (energy in/energy out) and its fuel use (typically expressed as heat rate in Btu/kWh). (Succar and Williams 2008).

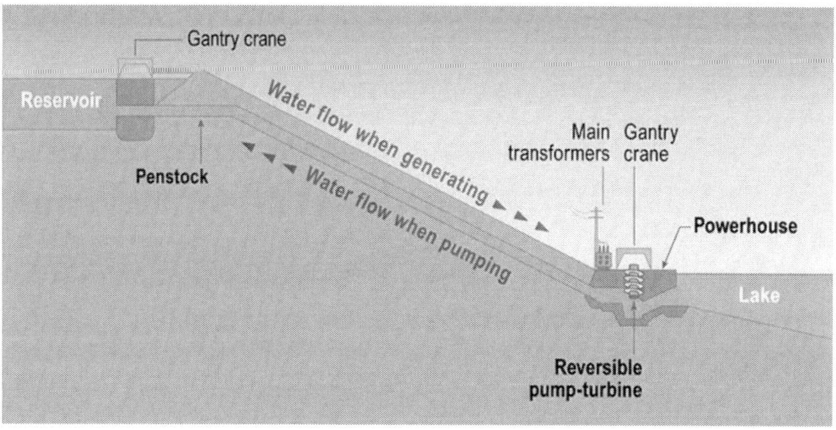

Figure 12.3. Simplified pumped-storage hydropower plant configuration.

The first CAES plant was completed in 1978 in Huntorf, Germany. It was designed primarily to provide "black start" (provide a source of power to start conventional generators after a system- wide failure), and it was rated at 290 MW with 2 hours of capacity (Crotogino et al. 2001). A second plant was built in 1991 in McIntosh, Alabama (Schalge and Mehta 1993). It has a rating of 110 MW for 26 hours, providing firm capacity and load-leveling services. Both plants inject air into underground caverns solution mined from salt formations (Succar and Williams 2008). This plant has a single turbo-machinery drive train using a common motor-generator set connected to the compressor and expander via clutches. This results in turnaround times from compression to expansion of approximately 30 minutes, limiting its use in providing operating reserves and other services requiring fast response.

Proposed CAES plants include a dedicated motor drive compressor and expander-generator that would eliminate the single turbo-machinery train (Norton Energy Storage 2000). This would allow for faster switchover from compression to generation, thus increasing its usefulness for providing ancillary services and responding to increased variability of net load. Once operating, CAES plants can provide rapid ramp rates; the McIntosh plant is capable of ramping at approximately 18 MW (16% of full output) per minute, or rates that are more than 50% greater than a typical gas turbine (Succar and Williams 2008).

Figure 12.4 shows a representative conceptual configuration of a CAES plant.

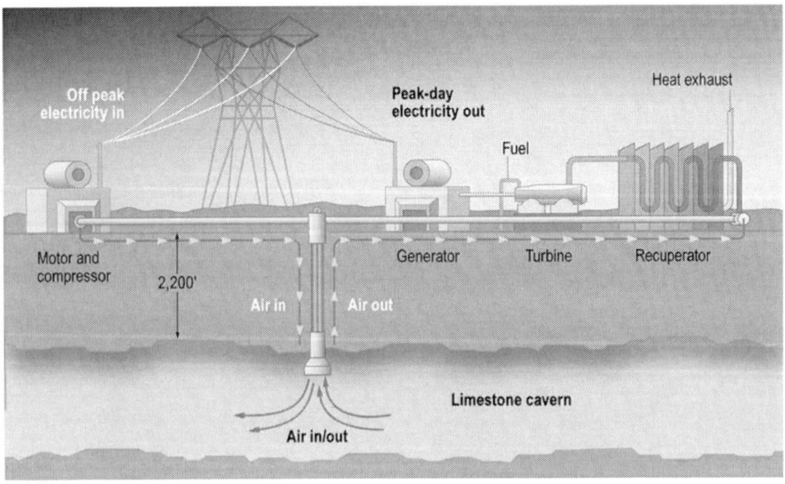

Figure 12.4. Configuration of a compressed air energy storage plant.

The large volume of air storage required for CAES is most economically provided by geological structures (Allen 1985; Korinek et al. 1991). The two existing CAES facilities use salt domes, where the cavity is formed by solution mining: fresh water is pumped into the formation to dissolve the salt, and brine is pumped to the surface for disposal or other use (Thoms and Gehle 2000). Domal salt formations are self-healing, meaning pores on the cavity walls seal themselves with available air moisture, virtually eliminating the possibility of air leakage.

Other proposed formations for CAES include bedded salt, which features thinner "layers" of salt. CAES can also potentially be deployed using aquifers, depleted natural gas formations, and hard- rock caverns. A variety of alternative and advanced CAES cycles have been proposed, and these are discussed in Section 12.1.4.3.

12.3.3. Technologies Not Included in RE Futures Scenario Analysis

The following technologies offer substantial potential benefits in many applications, but were not included in the Renewable Electricity Futures modeling as they either provide services not explicitly evaluated in the analysis or have not yet been significantly commercialized in grid storage applications.

12.3.3.1. Flywheels

Flywheels store energy in a rotating mass. Flywheels feature rapid response and high efficiency, making them well suited for frequency regulation. Several flywheel installations have been planned or deployed in locations where frequency regulation markets exist in the United States (Parfomak 2012).

12.3.3.2. Capacitors

Capacitors (including supercapacitors and ultracapacitors) are devices that store energy in an electric field between two electrodes (EPRI/DOE 2003). Capacitors have among the fastest response time of any energy storage device, and they are typically used in power quality applications such as providing transient voltage stability. However, their low energy capacity has restricted their use to short time-duration applications. A major research goal is to increase their energy density and increase their usefulness in the grid (and potentially in vehicle applications) (Hadjipaschalis et al. 2009).

12.3.3.3. Superconducting Magnetic Energy Storage

Superconducting magnetic energy storage (SMES) stores energy in a magnetic field in a coil of superconducting material. SMES is similar to capacitors in its ability to respond extremely fast, but it is limited by the total energy capacity. This has restricted SMES to "power" applications with extremely short discharge times (Luongo 1996; Feak 1997). Several demonstration projects have been deployed (Ali et al. 2010), and reducing costs by using high-temperature superconductors is a major research goal (Fagnard et al. 2006).

12.3.3.4. High-Power Batteries

High-power batteries are associated with the provision of contingency reserves, load following, and additional reserves for issues such as forecast uncertainty and unit commitment errors. This set of applications generally requires rapid response (in seconds to minutes) and discharge times in the range of up to approximately 1 hour.

These applications are generally associated with several battery technologies, which include lead-acid, nickel-cadmium, nickel-metal hydride, and (more recently) lithium-ion. With their rapid response, batteries can provide power quality services such as frequency regulation, but the continuous cycling requirement can limit life of current technologies (Peterson, Apt, and Whitacre 2010). Lithium-ion batteries are currently the primary candidate for large-scale deployment in battery electric vehicles (EVs) and *plug-in hybrid electric vehicles* (PHEVs), and improvements in batteries designed for vehicles could be applied to stationary applications (Wadia et al. 2011). Several demonstration projects have been built using these technologies to provide operating reserves. Details of cost and performance are provided in EPRI/DOE (2003) and EPRI (2010).

12.3.3.5. Electric Vehicles and the Role of Vehicle-to-Grid

EVs (used here to represent both "pure" electric vehicles and plug-in hybrid electric vehicles) are a potential source of flexibility for variable-generation applications. Charging of EVs can potentially be controlled and can provide a source of dispatchable demand and demand response. Controlled charging can be timed to periods of greatest variable-generation output, while charging rates can be controlled to provide contingency reserves or frequency regulation reserves. Vehicle-to-grid (V2G) (where EVs can partially discharge stored energy to the grid) may provide additional value by acting as a distributed source of storage. EVs could potentially provide all three grid services discussed previously. Most proposals for both controlled charging and V2G focus on short-term response services such as frequency regulation and contingency. Their ability to provide energy services is more limited by both the storage capacity of the battery and the high cost of battery cycling. This could restrict their ability to provide time shifting (energy arbitrage) beyond their ability to perform controlled charging.[168] The role of V2G is an active area of research, and because EVs in any form have yet to achieve significant market penetration, assessing their potential as a source of grid flexibility is difficult. However, analysis has demonstrated potential system benefits of both controlled charging and V2G (Denholm and Short 2006). The role of EVs as an enabling technology requires additional analysis of their unique temporal characteristics of availability, unknown battery costs and lifetimes, and the availability of smart charging stations to maximize their usefulness while parked.

12.3.3.6. Hydrogen Energy Storage and Fuel Production

A hydrogen energy storage system consists of an electrolyzer, storage tanks or underground cavern storage, and either a fuel cell[169] or combustion technology to produce electricity from hydrogen. Hydrogen has been produced industrially via electrolysis since the 1920s. There are currently no utility-scale installations using hydrogen as an energy storage medium; however, electrolyzers and fuel cells are commercially available, and electrolysis is used in a variety of industrial processes (Suresh et al. 2010).

Megawatt-scale hydrogen energy storage systems—using both above-ground storage (in tanks) and below-ground storage in formations similar to CAES—have been proposed (Kroposki et al. 2006). Because compressed hydrogen has a higher energy density than air, a storage cavern could store more energy in the form of hydrogen than could compressed air.

The primary disadvantages of hydrogen energy storage are the relatively low round-trip efficiency (between 28% and 40% depending on electrolyzer and fuel cell efficiencies) and the high cost of fuel cells and electrolyzers (Steward et al. 2009). Recent research has focused on cost reduction and efficiency improvements for fuel cells and electrolyzers, as well as on combining the electrolysis and fuel cell functions in a single "reversible" fuel cell device (Hauch et al. 2006; Milliken and Ruhl 2003; TMI 2001). This could increase efficiency and lower costs for hydrogen storage system (TIAX 2002).

A hydrogen energy storage facility could provide increased flexibility and unique revenue opportunities to utilities, which could sell or use the hydrogen for other applications. Hydrogen could be mixed with natural gas for additional flexibility in power generation from the storage system, but this has yet to be demonstrated on a commercial scale. The use of hydrogen as a transportation fuel represents a potentially large market (Greene et al. 2008). In addition to hydrogen, there are pathways to use electricity to produce liquid or gaseous fuels for vehicles or energy storage (Sterner 2009).

TEXT BOX 12.1. DEFINING THE COST OF ELECTRICITY STORAGE

A critical issue when discussing the costs of storage technologies is that storage devices in electric applications have both a power component (kW of discharge capacity) and an energy component (kWh of discharge capacity, which may also be expressed as hours of discharge at rated output). The total cost of a storage application must account for the ratings of both components, and it may be expressed differently depending on the application or audience. For example, because utilities universally define the cost of power plants only in terms of rated power ($/kW), they would expect to see costs in these terms, with the hours of storage (kWh capacity) expressed separately. A grid storage plant therefore might be expressed as costing $2,000/kW for a device with eight hours of discharge capacity. On the other hand, the battery community typically expresses costs in terms of rated energy ($/kWh), and it may or may not include the power component in the cost. So the cost of a battery might be stated as $500/kWh with the power capacity of the battery established separately. When evaluating the economics of storage technologies, care must, therefore, be taken to ensure that the costs for meeting both kW and kWh specifications are included and that both components are "sized" properly for any specific application.

12.3.4. Technology Cost and Performance

Limited deployment of many emerging energy storage technologies makes the estimation of costs challenging when deployed at scale. Even more mature technologies, such as PSH and CAES, have not been built in the United States in some time,[170] so the cost of the next plant is somewhat uncertain. Furthermore, PSH and CAES depend on site-specific geologic conditions, which make costs difficult to generalize. When considering costs of all storage technologies, the different applications must be considered. Storage technology costs include both an energy component and a power component, and the total cost of a storage device includes both components, within the limits of the target application. (This is discussed in more detail in Text Box 12.1.) Because the RE Futures modeling considered only bulk applications, only devices with multiple hours of discharge were evaluated. For uniform comparison, total costs were reported on a cost-per-kilowatt basis, where this cost includes both the power component and the energy component.

12.3.4.1. High-Energy Batteries

Present and future costs for many battery types are uncertain, particularly for flow batteries, due to the relative immaturity of the technology. Table 12.3 provides several estimates for the cost of several battery technologies providing energy services (with an energy capacity of at least 4 hours of continuous discharge).

Table 12.3. Battery Cost Estimates for Grid Storage Applications

Type	BOP[a] ($/kW)	Battery ($/kWh[b])	Storage Hours	Total/$/kW	Source
Vanadium	606	155–251	10	2,600–3,110	EPRI/DOE (2004)
Flow Battery (Several Technologies)	423–1,300	280–450	4	1,545–3,100	Rastler (2009)
NaS	450–550	350–400	4	1,850–2,150	Rastler (2009)
NaS	–	–	7.2	2,590	Nourai (2007)
Li-Ion	350–500	400–600	4	1,950–2,900	Rastler (2009)

[a] Balance-of-plant including power conversion system

[b] Although this column implies only the energy component, these estimates include the power component of the battery. As a result, the values in this table cannot be adjusted for more or less energy (hours of storage). Each cost assessment must be examined individually to determine the component costs.

Cost breakdowns for battery systems, including the balance of systems, installation, and other components, are provided by EPRI/DOE (2004) and Nourai (2007). The assumed cost for high- energy batteries (8–10 hours of discharge capacity) was $3,990/kW in 2010,[171] decreasing roughly linearly to $3,200/kW by 2050. Details about battery cost assumptions are provided in Black & Veatch (2012).

With battery efficiency, it is important to consider the alternating current (AC)-to-AC round-trip efficiency—battery efficiencies are often reported on a direct current (DC) basis without power conversion efficiencies—and to include the effect of "parasitic" loads, such as heating and cooling of batteries and power-conditioning equipment. Typical total AC-to-AC round-trip efficiencies for flow batteries and NaS are in the range of 65%–75%, including parasitic loads (Rastler 2008; Nourai 2007). Higher round-trip efficiencies for lithium-ion batteries have been reported in the range of 90% (KEMA 2008); however, this value does not include certain parasitic loads that can be considerable. A net roundtrip efficiency of 75% was assumed in this report.

12.3.4.2. Pumped-Storage Hydropower

Figure 12.5 provides historical cost data for U.S. PSH plants, inflated to 2009 dollars. There is a general trend toward increasing costs, with the last three plants constructed costing more than $1,000/kW.

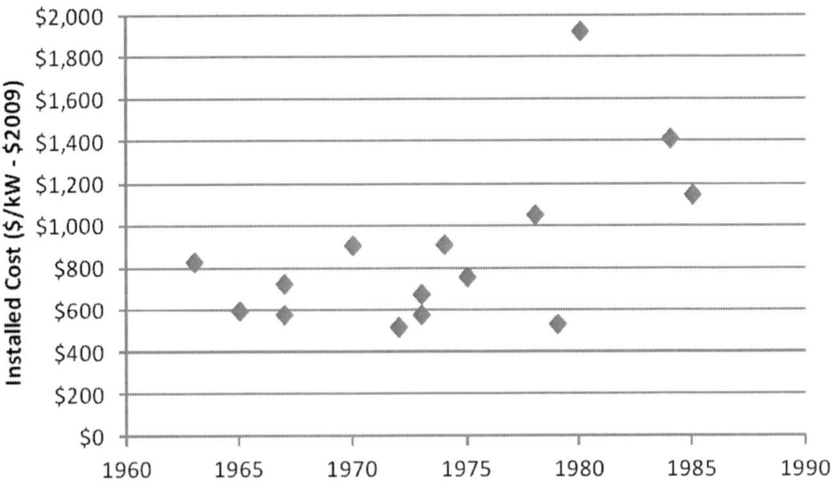

Figure 12.5. Installed cost of pumped-storage hydropower plants in United States.

Table 12.4. Recently Completed or Proposed Pumped-Storage Hydropower Plants[a]

Location	Plant Name	Capacity (MW)	$/kW	Date of Completion
United States				
California	Eagle Mountain	1,300	1,019	Proposed
California	Iowa Hills PS	400	1,344	Proposed
California	Lake Elsinore	500	1,500	Proposed
California	Red Mountain	900	1,900–2,100	Proposed
Utah	North Eden PS	700	1,011	Proposed
Utah	Parker Knoll PS	800	1,215	Proposed
Austria	Feldsee	140	750	2009
Austria	Reisseck_II	430	1,091	2008
Germany	Goldisthal	1,060	1,321	2003
Slovenia	Avce	180	711	2009

[a] This represents a small subset of the proposed plants in the United States.

The cost of new PSH plants will vary. The geotechnical and geological characteristics and complexity of site are major factors in PSH development costs. Typically, the largest costs are for development of a project's upper and lower reservoirs and for underground components. One example is the Helms pumped hydropower plant, which was completed in 1984 at a cost of $1,411/kW (2009 dollars), with approximately 50% of the cost being the reservoir, and 28% being the powerhouse (ASCE 1993). No large projects have recently been built in the United States; however, a number of projects have been completed worldwide in the last decade, and there are a significant number of proposed plants both in the United States and internationally.

Table 12.4 lists several recently completed plants in Europe (Deane et al. 2010), along with proposed plants in the United States; capital costs (in dollars-per-kilowatt) are adjusted to 2009 dollars (NWPCC 2008). There are also a large number of proposed plants in Europe, with costs estimated in the range of $700/kW to more than $3,000/kW.

Deane et al. (2010) provides a more comprehensive discussion of recent and projected future costs. Recent engineering estimates of new PSH construction costs per kilowatt in the United States include $2,100–$4,000 (Rastler 2009), $2,000–$4,000 (Black & Veatch 2012), and $5,595 (EIA 2010). A large component of this very large range is due to the variation in local conditions—low-price estimates may assume the availability of existing reservoirs (including abandoned mines or other formations), while the high estimates may assume "green field" development or modification of both reservoirs. Generating a supply curve would require evaluation of each individual potential site. Efforts have been initiated to characterize potential new PSH development at scale, but additional data were unavailable at the time of this analysis.

As a result, cost estimates were based on a combination of proposed plant costs described above and engineering estimates, focusing on lower-cost PSH opportunities. Two cost points were identified, at $1,500/kW and $2,000/kW.

One of the primary challenges associated with PSH development is the long construction time, as well as associated risks and uncertainty. State and local application and permitting (including obtaining water rights), FERC permitting, and construction require 10–12 years based on current schedules. Closed-cycle plants could reduce licensing and construction times to 6–8 years. These times (and resulting costs) can be increased due to siting opposition and environmental regulations (Strauss 1991).

Existing PSH facilities in the United States—most of which were constructed during the 1960s, 1970s, and 1980s—have high availability and few forced outages. The great majority of U.S. plants have multiple reversible pump-turbine motor-generator units. Reversible units operate as a motor and pump in the "pumping" mode, and as a turbine and generator in the "generating" mode. Having multiple units per plant allows for scheduling maintenance on one unit while keeping the other units available, typically minimizing effects on overall plant availability.

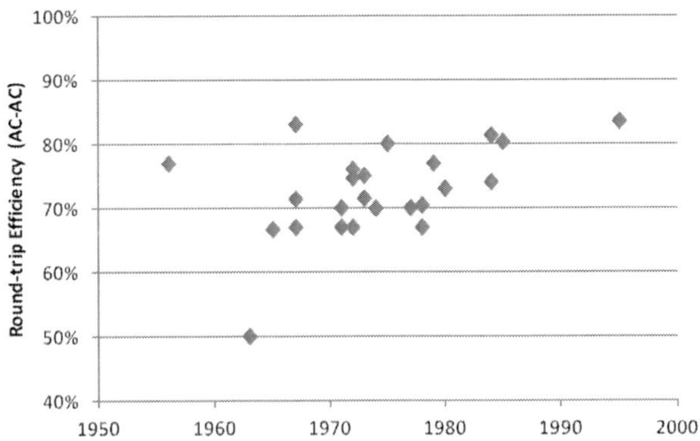

Source: Performak 2012.

Figure 12.6. Historical efficiencies for pumped-storage hydropower plants in United States.

Figure 12.6 provides the round-trip efficiencies for existing U.S. PSH plants. There has been a trend toward increased efficiencies, and proposed plants have efficiencies that exceed 80% on an AC-to-AC basis (ASCE 1993). Assumed efficiency for new PSH for this study was 80%. There is little loss of performance due to age or throughput. Plants are upgraded through efficiency improvements and life extension on a project-by-project basis, and most U.S. projects have been modernized through runner (turbine) replacements, generator rewinds, control system upgrades, and other incremental improvements. Lifetimes of PSH plants can exceed 60 years (ASCE 1993).

Older PSH plants can require up to 30 minutes to switch between pumping and generation. However, modern PSH plants enable fast ramping rates in both pumping and generation modes and can begin pumping or generating within seconds.

RE Futures assumed that new PSH deployments would include variable speed (also referred to as "adjustable speed") operation. This technology has not yet been applied in a major U.S. installation, but has been used in several international plants (Yasuda 2000). Among the benefits of variable speed operation are faster response to grid requirements, higher efficiencies, ability to accommodate greater ranges of "head," and wider unit and plant operating ranges (i.e., an ability to operate with a lower minimum load in megawatts).

12.3.4.3. Compressed Air Energy Storage

The cost of CAES plants is driven by aboveground components, including compressors and the expander/generator equipment, as well as by belowground components. Aboveground equipment components are based largely on standard components, with the uncertainty in cost based largely on large swings in commodity prices and the general cost of capital-intensive projects. The largest uncertainty associated with CAES is related to underground cavern development and is especially associated with unproven approaches such as development in bedded salt and aquifers.

Salt caverns are generally the most economical excavated formations for siting CAES plants. Excavation costs for salt caverns, which are constructed by solution mining, can be kept extremely low compared to the costs for bedded salt formations, aquifers, and hard rock mining. Based on current experience with the construction of natural gas storage reservoirs and the Big Hill strategic petroleum reserves in Texas, costs can be maintained at approximately $\$2/m^3$ of excavated cavern for solution mining compared to $\$20/m^3$ in aquifers, and $\$300/m^3$ in hard rock granite.

Table 12.5 provides several cost and performance estimates for proposed CAES plants. Table 12.6 breaks down costs for a conventional CAES system deployed with a salt cavern.

Table 12.5. Cost and Performance Estimates for Four Proposed Compressed Air Energy Storage Plants[a]

Name	Location	Cavern Type	Capacity (MW)	Cost ($/kW)	Heat Rate (Btu/kWh)	Energy Ratio[b]
Iowa Stored Energy Park	Dallas Center, Iowa	Aquifer	–	933–1,014	4,420	0.77–0.89
Norton Energy Storage	Norton, Ohio	Depleted hard-rock mine	2,700	–	3,860–4,300	0.7
PG&E	Kern County, California	Porous rock	300	1,187	–	–
Seneca (NYSEG/ Iberdrola)	Schuyler County, NY	Bedded salt	150	833	–	–

[a] Performak 2012

[b] The energy ratio is defined as the amount of electrical energy in per unit of generation. Note that this number is less than 1 because CAES is a hybrid system that uses natural gas. The efficiency of a conventional CAES plant cannot be easily defined as a single number because it uses two different energy sources.

Table 12.6. Cost Breakdown for a Conventional Compressed Air Energy Storage System Deployed in a Salt Cavern

Component	Cost ($/kW)	Fraction of Total
Compressor	87	11%
Heat exchanger	34	4%
High pressure expander	62	8%
Low pressure expander	144	19%
Electrical	45	6%
Construction, labor, indirect costs	324	42%
Cavern development	77	10%
Total	774	100%

Source: CEC 2008.

For RE Futures, the aboveground costs were based on a "reference plant" with a capacity of 220 MW. This reference plant assumes a multi-stage compressor, with the first stage using an axial flow compressor with a discharge pressure of 160 pounds-force per square inch gauge (psig) and requiring a power input of 90 MW. The discharge air is passed through an intercooler, which reduces the air's specific volume and temperature in preparation for the second stage of the compression process in which the air is compressed to its final storage pressure of 1,250 psig.

Three installed costs were assumed for new CAES development for RE Futures: $900/kW for deployment with salt domes, $1,050/kW in bedded salt, and $1,200/kW in aquifers. These values are based on engineering estimates, discussed in detail in Black & Veatch (2012), and are within the range cost estimates in Table 12.5 of $730/kW to $1,200/kW for deployment in salt and aquifers. Hard rock caverns that must be excavated were not included in RE Futures, although opportunities for CAES deployments exist in depleted mines.

RE Futures assumed a CAES energy ratio of 0.8 kWh_{in}/kWh_{out} and a heat rate of 4,910 Btu/kWh. These estimates were based on expected performance of the proposed (and subsequently cancelled) Iowa Stored Energy Park (Black & Veatch 2005; Schulte et al. 2012). The reference plant for RE Futures assumed dedicated motor and generators to allow fast switchover times and provision of operating reserves. RE Futures assumed a very high availability, based on both the similarity of CAES to natural gas turbines and the historical performance of the McIntosh Power Plant in Alabama. Plant lifetimes are expected to be similar to conventional gas turbine plants, typically exceeding 20 years with normal maintenance (Crotogino et al. 2001). Additional discussion of CAES cost and performance assumptions is provided in Black & Veatch (2012).

12.3.5. Technology Advancement Potential

12.3.5.1. Batteries

There is considerable opportunity for cost reduction and improvements in many battery technologies. EPRI/DOE (2003 and 2004) describe several cost reductions that could result from engineering and manufacturing scale-up of flow batteries and NaS batteries. Historical "learning curves" show continued progress of both "mature" battery technologies and newer technologies such as lithium-ion. Figure 12.7 and Figure 12.8 illustrate the historical increases in energy density as well as cost for a variety of energy storage devices.

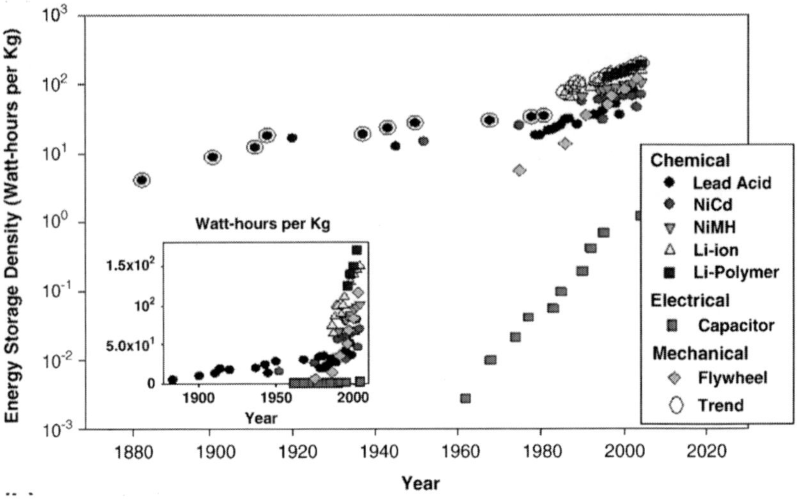

Source: Koh and Magee 2008.

Figure 12.7. Historical improvements in storage energy density.

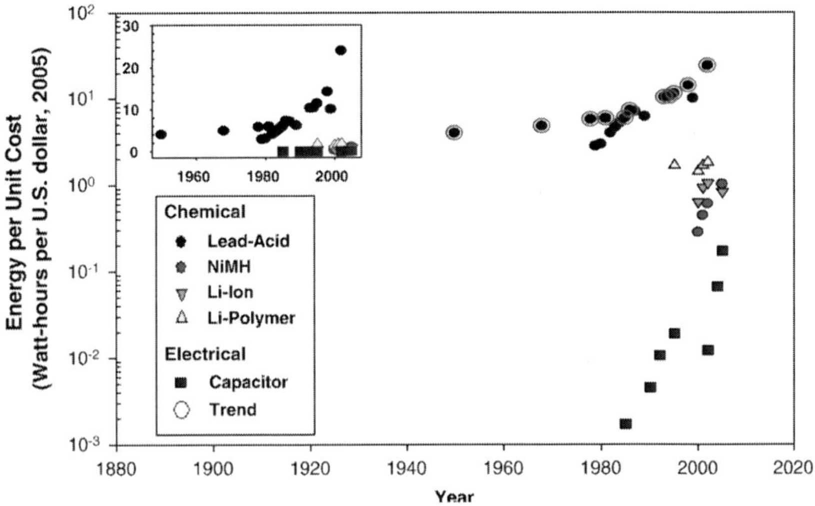

Source: Koh and Magee 2008.

Figure 12.8. Historical improvements in energy storage cost.

The emergence of nano-scale science provides opportunities for entirely new battery structures that could dramatically improve the power and energy density of several types of batteries. DOE (2007) provided a detailed discussion of the potential opportunities for batteries. The target for the Advanced Research Projects Agency-Energy (ARPA-E) stationary storage program is $100/kWh.[172] In addition to research on stationary batteries, efforts to reduce the cost of transportation batteries could have significant impact on their application for grid services. RE Futures did not consider the impact of fundamental breakthroughs in battery science on reduced costs and subsequent deployment, nor did it evaluate the distribution level benefits of battery deployment.

12.3.5.2. Pumped-Storage Hydropower

Pumped-storage hydropower is considered a mature technology. However, incremental improvements in efficiency are possible, and the flexibility of existing and future plants may be improved using variable speed drive technologies. Other possible developments include use of saltwater PSH facilities in coastal regions and underground PSH (Tanaka 2000). Resource availability or detailed cost estimates of these alternative configurations were not available, so they were not considered for RE Futures.

12.3.5.3. Compressed Air Energy Storage

Although CAES is based on mature technologies, there are several possible advancements in conventional CAES. Previous CAES plants used components that were not optimized for the unique characteristics of the CAES expansion cycle. This is partially due to the small market for which developing dedicated equipment would not be worthwhile. A large CAES market could drive development of custom turbo-machinery, improving the efficiency of CAES components. Alternatively, several proposed CAES configurations use standard combustion turbines, potentially lowering cost significantly (Nakhamkin 2008). At least one proposed plant has considered an advanced CAES cycle (NYSEG 2009; Rettberg 2010).

Several other advanced CAES concepts were not included in RE Futures. These include aboveground CAES using pipes or other containers (which would have only a few hours of storage) or alternative fuels (such as liquid or gas biofuels). Other configurations not included in RE Futures include several proposed concepts that do not require natural gas. These include adiabatic CAES, which stores the heat of compression and uses this stored energy during expansion. This type of configuration has yet to be constructed, with cost and performance estimates based only on engineering studies (Grazzini and Milazzo 2008). However, at least one demonstration plant has been proposed in Europe (RWE 2010). Another approach being explored is isothermal CAES, which maintains constant temperature (Kepshire 2010).

12.4. Resource Cost Curves

12.4.1. Batteries

Batteries do not have the geologic constraints of CAES or PSH. They also do not have fuel or water requirements, so they were assumed to be deployable at scale within each region.

12.4.2. Pumped-Storage Hydropower

New PSH development requires sufficient land for construction of the two requisite reservoirs, with a sufficient elevation difference between the reservoirs to enable economical generation.

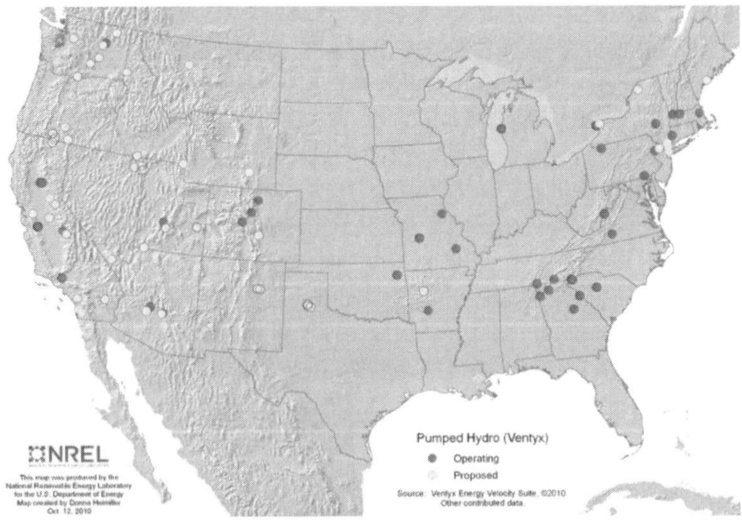

Figure 12.9. Location of existing and proposed (with Federal Energy Regulatory Commission preliminary permits) pumped-storage hydropower installations in the contiguous United States.

Many areas of the United States offer suitable topography, and the technical potential of PSH is extremely large. Although there is no recent comprehensive estimate of PSH potential, older studies indicate the availability of hundreds of conventional PSH sites, more than 1000 GW of potential capacity in just six western states (Allen 1977), and more than 100 GW of potential in the Eastern Interconnection (Dames and Moore 1981). This capacity is roughly equivalent to the installed generation capacity for all of the United States (EIA n.d.). These older assessments include some areas that would be very difficult (or impossible) to develop based on current environmental restrictions. However, the capacity of recently proposed plants (exceeding 40 GW) is greater than the existing installed U.S. storage capacity and suggests there are considerable opportunities for new PSH capacity. RE Futures used an estimate for PSH availability based solely on the location and sizes of proposed plants for which data could be obtained (FERC n.d.). As a result, the developable potential of new PSH was fixed at 35 GW. Although this is much smaller than the technical potential of more than 1,000 GW, there are no data to estimate current development costs of this potential beyond engineering estimates that are as high as $5,595/kW. (Cost estimates are actually provided for much of this potential in the original assessment documents from the 1970s, but these costs are unlikely to reflect current market conditions.) The 35 GW of proposed capacity likely represents lower-cost opportunities as reflected in proposed costs, and reviews of these proposals were used to generate the two price points of $1,500/kW and $2,000/kW discussed in Section 12.1.3.2. Based on the reviews of proposed plants, the lower-cost value ($1,500/kW) was assigned to 10 GW of potential, while the higher cost ($2,000/kW) was assigned to 25 GW of potential. Figure 9 provides a map of the existing and proposed plants in the United States. The proposed plants were used to create a supply curve for new development (Figure 12.10), with the two cost points spread uniformly across the resource. Overall, the fact that costs could only be assigned to less than 4% of the technical potential indicates a fundamental need for understanding the potential of new PSH development.

12.4.3. Compressed Air Energy Storage

Estimating the amount of underground formations available for CAES is very difficult. Some estimates indicate that more than 75% of the land area of the United States could provide suitable geology for CAES projects (Allen

1985; Mehta 1992). However, each potential site must be individually screened, and this has proved challenging. For RE Futures, CAES deployment was limited to three options: domal salt, bedded salt, and porous rock (primarily aquifers).

Aquifer storage caverns are composed of permeable or fractured rock, and these formations are currently used to store natural gas. The identification of the necessary rock types and formations requires extensive geological testing to ensure the appropriate conditions exist for storage of compressed air. The major criteria for successful aquifer storage caverns are:

1. The existence of a structure shaped like an inverted saucer with the capability of sufficient air storage volume, which is determined from the porosity of the porous media comprising the aquifer
2. A continuous impermeable overlying caprock with a low permeability that inhibits the stored pressurized air from displacing water contained within the caprock pores
3. Sufficient structure depth (at least 600–800 feet or 183–244 m) having the full hydraulic pressure to assure adequate capacity of the aquifer pore volume along with the required characteristics to ensure adequate airflow from the formation
4. Permeability of the storage zone, not only in the air reservoir but also in the aquifer surrounding the structure.

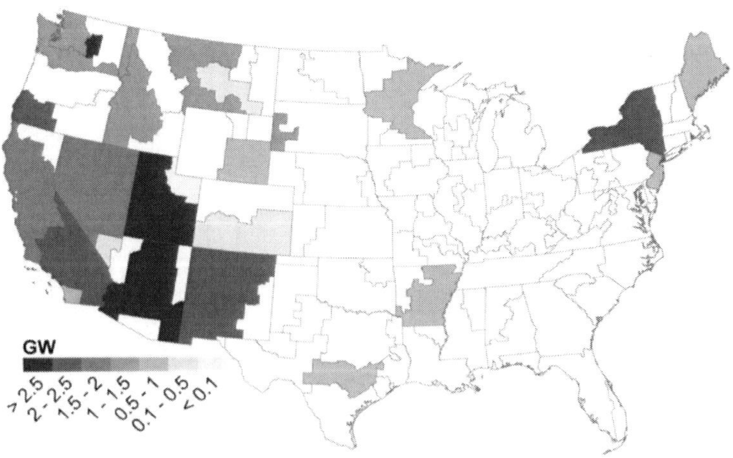

Figure 12.10. Pumped-storage hydropower resource potential used in the ReEDS modeling.

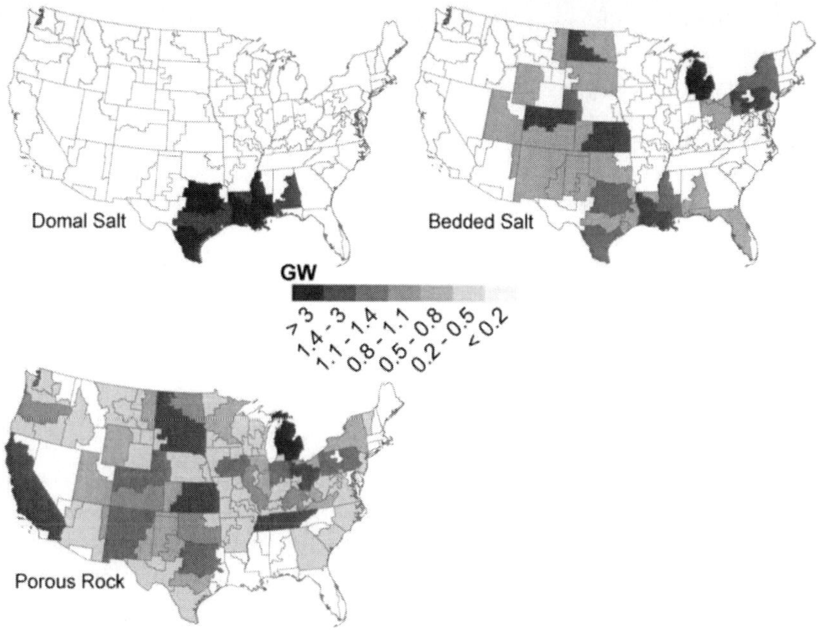

Figure 12.11. Assumed availability of compressed air energy storage in domal salt ($900/kW), bedded salt ($1,050/kW), and porous rock ($1,200/kW).

The air under pressure will displace the water in the structure to form the storage reservoir. High permeability is needed to give a reasonable time to develop the reservoir and maintain proper airflow during injection and withdrawal.

CAES was excluded in certain porous rock formations such as depleted gas wells, except in California, where this application has been examined in some detail, and there is at least one proposed plant (Hobson et al. 1977; CEC 2008). Use of CAES in hard rock was also excluded due to lack of data. Although the cost of excavating hard rock solely for use in CAES is typically considered cost prohibitive, CAES could be used in existing depleted hard rock mines, and at least one large (2,700-MW) CAES plant has been proposed used an existing hard rock mine (Bauer and Webb 2000).

Figure 12.11 provides the estimates of CAES availability (in gigawatts) for the locations (by ReEDS balancing area), the availability (in gigawatts), and assumed cost (in dollars per kilowatt) for each of the three CAES deployment options (with the cost including both the power components and cavern development, assuming about 15 hours of storage capacity). For the contiguous United States, the potential CAES resource was estimated to exceed 120 GW, with about 23 GW in domal salt, 37 GW in bedded salt, and 62 GW in porous rock. No technology- driven cost improvements for CAES are assumed in the model scenarios.

12.5. Output Characteristics and Grid Service Possibilities

Output characteristics and grid service possibilities are discussed in Section 12.3.

12.6. Deployment in RE Futures Scenarios

Deployment of new storage capacity is observed in all model scenarios described in Volume 1, and greater storage deployment is realized in scenarios with greater levels of renewables, and particularly variable renewable, penetration. For the (low-demand) core 80% RE scenarios described in Volume 1, 80–131 GW of new storage capacity was installed by 2050 in addition to the 20 GW of existing (PSH) storage capacity. Of the six core 80% RE scenarios, the constrained flexibility scenario projected the greatest level of storage deployment (152 GW of installed storage capacity by 2050). The constrained flexibility scenario was designed to capture greater institutional and technical barriers to managing variable generation, compared to the other 80% RE scenarios modeled. These barriers were implemented in ReEDS by halving the statistically calculated capacity values for wind and PV, increasing the reserve requirements for wind and PV forecast errors, reducing the flexibility of coal and biomass plants, and limiting the availability of demand response.[173] In the constrained flexibility scenario, new storage additions occur predominantly in the first two decades (2010–2030) of the study period, with an average annual installation rate of approximately 5 GW/yr and decade-averaged annual capital investments ranging from $4 billion/yr to $11 billion/yr between 2010 and 2030.[174] Figure 12.12 summarizes storage deployment in the constrained flexibility scenario, and Figure 12.13 shows the locations of storage deployment in the same scenario.

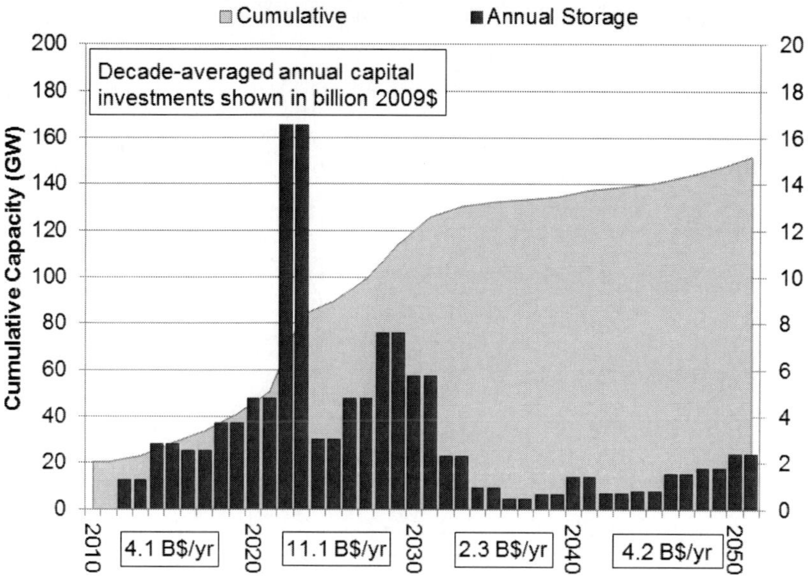

Figure 12.12. Deployment of energy storage technologies in the constrained flexibility scenario.

As discussed earlier, the modeled deployment indicates the general amount of storage that might be used to enable a high renewables scenario rather than to indicate a prescribed amount of each technology type. As a result of the modeling assumptions, most of the new storage is CAES; however, the tradeoff between CAES and PSH is largely due to the modeling and data limitations associated with the vast majority of potential PSH in much of the United States. In addition, the relative risk associated with CAES versus PSH was not considered. PSH is a proven technology, while CAES has yet to be deployed in either bedded salt or in porous rock formations, which represents a large fraction of assumed deployments. The limited deployment of batteries is due to their high cost and assumed minimal cost reduction but also to a lack of valuation of their benefits to the distribution system. This demonstrates an obvious discrepancy with relative historical and proposed deployment of these technologies, where PSH dominates. *The analysis of energy storage technologies for RE Futures demonstrates the need for more comprehensive estimates of the cost and resource availability for both CAES and PSH.*

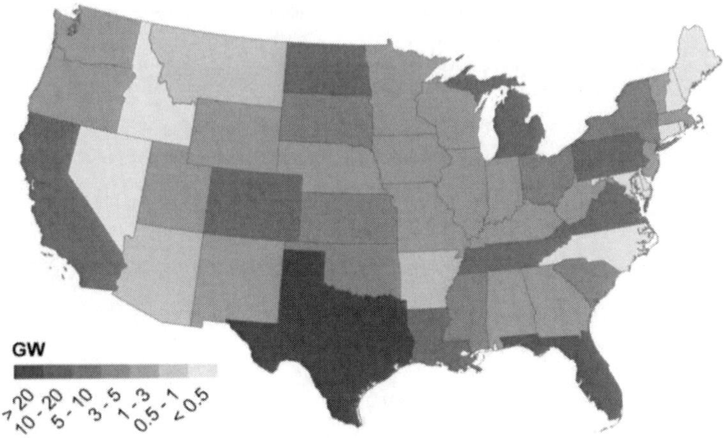

Figure 12.13. Regional deployment of storage in the contiguous United States in the constrained flexibility scenario.

Table 12.7 and Figure 12.14 show the variation in storage deployment between the low-demand core 80% RE scenarios and the high-demand 80% RE scenario. Between these scenarios, the 2050 installed storage capacity ranged from about 100 GW to 152 GW. A lower level of storage deployment is found under the 80% RE-ETI scenario, which included high levels of deployment of CSP with thermal storage and a corresponding lower deployment of variable generation technologies, thereby mitigating some of the need for the non-thermal storage technologies. Conversely, greater wind deployment in the 80% RE-NTI scenario and greater wind and PV deployment in the high-demand 80% RE scenario motivated high levels of storage deployment, although these two scenarios still realized slightly lower levels of deployment than the constrained flexibility scenario detailed above. Descriptions and results of the model scenarios are detailed in Volume 1.

12.7. Large-Scale Production and Deployment Issues

12.7.1. Environmental and Social Impacts

The impacts of energy storage are a function of two components. First is the localized impact due to development and direct use of the individual energy storage technologies. These vary significantly given the large differences in technology types. The second is associated with the upstream source of electricity, and the increased generation typically required due to inefficiencies in the storage process.

12.7.1.1. Land Use

Land use estimates for batteries are limited due to the lack of deployment at scale. For NaS, one estimate is approximately 211 m^2/MW with a 7.2-hour storage capacity (NGK n.d.), or approximately 300–350 m^2/MW for a 10- to 12-hour device more comparable to CAES or PSH. An estimate for a proposed (and subsequently cancelled) large (12 MW, 100–120 MWh) flow battery was approximately 850 m^2/MW (EPRI/DOE 2003) with additional land surrounding the facility (TVA 2001).

Table 12.7. Deployment of Energy Storage Technologies in 2050 under 80% RE Scenarios[a,b]

Scenario	Capacity (GW)
Constrained Flexibility	152
80% RE-NTI	142
High-Demand 80% RE	136
Constrained Resources	131
Constrained Transmission	129
80% RE-ITI	122
80% RE-ETI	100

[a] See Volume 1 for a detailed description of each RE Futures scenario.
[b] Capacity totals represent the cumulative installed capacity for each scenario.

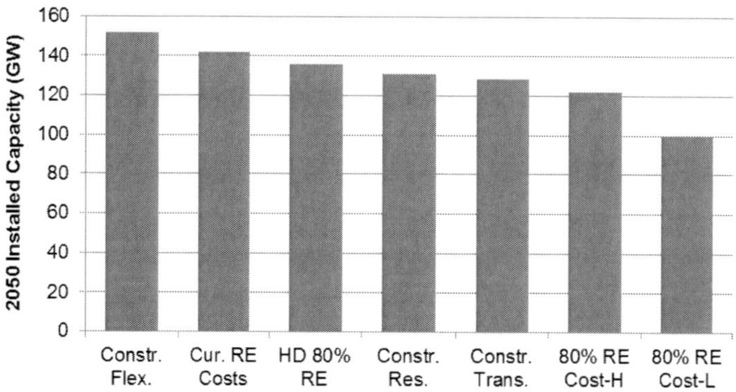

Figure 12.14. Deployment of energy storage technologies in 80% RE scenarios.

Land use impacts of CAES deployment are minimal because most of the plant is effectively underground. The land area estimates for one proposed CAES facility is approximately 140 m^2/MW (Norton Energy Storage 2000).

Pumped-storage hydropower can require a significant amount of land area for the upper and lower reservoir, depending on configuration. The total flooded area of three of the more recently constructed large PSH plants in the United States (the Bad Creek Hydroelectric Station in South Carolina, the Balsam Meadow Pumped Storage Project in California, and the Bath County Pumped Storage Station in Virginia) is in the range of 1,200 m^2/MW to 1,500 m^2/MW (ASCE 1993). Older PSH facilities with constructed upper and lower reservoirs have flooded areas that exceed 4,000 m^2/MW. New plants are more likely to have land use requirements towards the lower range, such as the proposed Eagle Mountain and Iowa Hill plants with flooded area requirements of approximately 1,100 m^2/MW (Tam 2008; Parfomak 2012). Additional discussion of land use associated with hydropower in general is provided in Chapter 8.

12.7.1.2. Water Use

For CAES, the dominant use of water is for formation of underground caverns in domal or bedded salt. Water use for solution mining is likely to be about 8 m^3 of water for each cubic meter excavated (Smith 2008) or about 4.8 million m^3 of fresh water withdrawals and brine management per 220-MW plant. Disposal of brine has been raised as a concern for some locations (Smith 2008). Additional cooling water is required during operation of the compressors, with one estimate of 2.5–3.0 million gallons per day for a 2700-MW facility (Ohio Power Siting Board 2001). Assuming a capacity factor of 25%, this corresponds to approximately 0.2 gallons/kWh.

Analysis and discussion of water impacts of PSH include Clugston (1980) and U.S. Bureau of Reclamation et al. (1993). Impacts on water quality and aquatic life have greatly delayed and even prevented operation of completed PSH facilities (Southeastern Power Administration 2009; U.S. GAO 1996). The actual water use and impacts of PSH depend partially on the source for the lower reservoir. Most existing U.S. PSH plants are "open-cycle" plants; that is, they use an existing water body, usually the lower reservoir, for one of their reservoirs. However "closed- cycle" plants—plants where both lower and upper reservoirs are constructed—will likely become more prevalent in the future because they minimize environmental effects as they do not interact with natural water bodies and they have little or no impact on aquatic life. Water sources for closed-cycle plants vary. Some proposed plants will use groundwater for the initial fill and make-up water required to replace seepage and evaporation. One estimate for make-up water for a 1,300-MW facility is 782 million gallons/yr (Tam 2008). Assuming a capacity factor of 25% (2,847 GWh/yr), this corresponds to a water consumption rate of approximately 0.3 gallons/kWh. At least one facility has proposed to use recycled wastewater, and it has been suggested that this could be a significant opportunity for other new PSH facilities (Yang and Jackson 2011).

12.7.1.3. Life Cycle Greenhouse Gas Emissions

Energy storage can add to net greenhouse gas (GHG) emissions in three ways. First, the losses associated with storage efficiencies increase the electricity needed to produce a unit of delivered energy via storage (energy storage losses can be partially offset by increased efficiency of thermal generators that is due to either operation that is closer to the "design point" or a reduced need for ancillary services [Denholm and Holloway 2005]). Second, energy storage technologies produce life cycle emissions that are due to construction and operations. These life cycle values for PSH, several battery types, and CAES (excluding natural gas use) are in the range of 5–40 grams equivalent carbon dioxide per kilowatt-hour (gCO_2e/kw) depending on operation, lifetime, and other factors (Denholm and Kulcinski 2004). This includes the methane emissions from vegetation decomposition by land flooded by new PSH reservoirs, which are relatively small, especially for sites in the United States (Gagnon and van de Vate 1997; Rosa and dos Santos 2000). Finally, CAES burns natural gas, emitting GHG emissions at a rate of about 215–240 gCO_2e/kWh of delivered energy, assuming a heat rate range of 4,000–4,400 Btu/kWh (plus GHG emissions associated with production and transport of natural gas.)

Given the uncertainty in storage technology mixes, and given limited data, the life cycle GHG emissions impacts due to energy storage manufacturing were not evaluated, resulting in a small underestimation of system-wide GHG emissions for the non-fuel storage component. However, the CAES fuel combustion emissions were counted. Thus, the degree of underestimation is likely very small because of both the limited deployment of storage and their relatively small emissions.

12.7.1.4. Other Waste and Emissions

In general, with the exception of CAES, energy storage does not require direct fuel or combustion processes, so it produces no direct air emissions. The use of natural gas in CAES produces the various impacts associated with gas exploration, production, transmission, and combustion. This produces emissions such as nitrogen oxides in a manner similar to conventional gas turbines, but at a correspondingly lower rate given the much lower heat rate. Nitrogen oxide emissions can be controlled using conventional emissions controls such as selective catalytic reduction, which has been proposed for use in the CAES plants under consideration (Norton Energy Storage 2000; CEC 2008.)

Batteries use a variety of materials, some of which are toxic. Lead and cadmium are examples, and collection and recycling programs are generally in place to avoid improper disposal (EPRI/DOE 2003). Additional programs would be required for new battery chemistries, depending on their level of deployment and materials used.

12.7.2. Manufacturing and Deployment Challenges

Both CAES and PSH are based on mature technologies that have been previously deployed in the United States at scale. For example, the equipment required for CAES is very similar to conventional gas turbines, and the historical installation of gas turbines has exceeded 10 GW/yr in some years (EIA n.d.). An additional discussion of issues related to PSH manufacturing is provided in Chapter 8. For batteries, the primary issues for large-scale deployment may be related to a combination of materials requirements and competition with automotive applications. Wadia et al. (2011) discusses this issue at length and finds essentially no material challenges for some technologies such as NaS, but potential constraints on others, such as Vanadium Redox or certain lithium-ion batteries using cobalt.

12.8. Barriers to High Penetration and Representative Responses

Although capital cost is a primary barrier to deployment of energy storage, many regulatory and market barriers prevent energy storage from competing equally with more conventional technologies that provide energy and capacity services.

Table 12.8. Barriers to High Penetration of Electricity Storage Technologies and Representative Responses

R&D	Barrier	Representative Responses
Batteries	High capital cost, limited cycle life	Conduct fundamental science and engineering to improve power and energy density; research new electrolyte materials; standardize and integrate power conversion systems
CAES	Cost, efficiency, unproven availability of sites	Research and development into advanced CAES cycles, including cycles that reduce or eliminate use of natural gas; demonstrate CAES in aquifers, bedded salt, and depleted gas wells; conduct detailed national screening of suitable geologic formations
PSH	Availability of sites	Conduct detailed national screening of suitable formations
Market and Regulatory	**Barrier**	**Representative Responses**
All	Limited value proposition for energy storage	Provide comprehensive analysis of the system benefits of storage, including utility operations models that accurately represent the complete set of benefits of energy storage over multiple timescales
All	Unclear treatment of energy storage in regulatory framework	Establish a regulatory framework that provides fair and equitable cost-recovery mechanisms for new storage development congruent with its system benefits
Environmental and Siting	**Barrier**	**Representative Responses**
PSH	Land and water use	Conduct detailed screening of opportunities for closed- cycle plants, and siting on brown fields and other disturbed land

Table 12.8 summarizes actions that could enable greater use of energy storage. Table 12.8 includes only a small subset of energy storage technologies. Other existing and emerging storage technologies could be deployed in substantial numbers given appropriate decreases in costs.

12.8.1. Research, Development, and Deployment

For batteries (and other electro-chemical storage technologies), most RD&D efforts are focused on reducing capital cost, increasing power and energy density, and increasing lifetimes. Several recent reports identify fundamental research and engineering needs for improving basic technologies, as well as developing manufacturing techniques to bring laboratory technologies to commercial products and to bring next-generation technologies to market (Hall and Bain 2008; APS 2007; DOE 2007).

The primary RD&D issues associated with both PSH and CAES are related to resource assessment. There is no known comprehensive assessment of the total availability of PSH or CAES geology to assess the resource potential, although efforts are underway by DOE and others to perform additional resource assessment for both technologies (Rogers et al. 2010). Additional near-term RD&D activities can aid in developing dedicated turbo-machinery equipment for CAES, providing incremental improvements in both cost and performance if deployed at large scale. Similarly, RD&D can provide incremental improvements to PSH pump- turbine equipment, and could examine opportunities to convert existing single speed units to variable speed operation (ORNL et al. 2010).

12.8.2. Market and Regulatory

The primary market and regulatory barrier to storage deployment in general is lack of appropriate valuation of storage benefits. Until recently, the value of ancillary services was largely unquantified. The creation of wholesale markets has placed value on those services and has increased participation of energy storage devices, but the level of participation varies by market.[175] In 2007, FERC issued Order 890 requiring wholesale markets to consider non-generation resources for grid services (Kaplan 2009). Since then, independent system operators and regional transmission operators have increased market access, including creating new tariffs for energy storage, and several storage projects have been proposed or built to take advantage of high-value ancillary service markets. However, market rules are still evolving in some locations (and of course, much of the United States has no access to restructured energy markets). A main benefit of energy storage is also its ability to provide multiple services, including load leveling (and associated benefits such as a reduction in cycling-induced maintenance) (Troy et al. 2010; Grimsrud et al.

2003) along with regulation and contingency reserves and firm capacity (Eyer and Corey 2010). However, quantifying these various value streams is difficult without sophisticated modeling and simulation methods. Because the economic analysis is difficult and benefits of storage are often uncertain, utilities tend to rely on more traditional generation assets, especially in regulated utilities where risk is minimized and new technologies are adopted relatively slowly. Changing and uncertain regulations and market structures also deter projects with long development times such as PSH, or uncertain technology challenges, such as CAES with site-specific geological screening requirements.

There are additional barriers to individual technologies. For PSH, the challenge of long permitting times could be reduced by applying an alternative licensing process to closed-cycle plants. These plants could be candidates for a streamlined FERC permitting process given their lack of interaction with any active stream, lake, or estuary.

12.8.3. Siting and Environmental Barriers

The primary siting challenge for new PSH and CAES is finding suitable geologic formations, discussed previously. PSH also faces potential opposition due to environmental impacts, which can be partially mitigated using closed-cycle plants. Both PSH and CAES plants are typically large, requiring new high-voltage transmission, which adds additional challenges, especially considering potentially remote locations. For batteries, the primary concern is the potential release of materials from liquid electrolyte flow-batteries. Proper containment and mitigation is required to minimize possible impacts (TVA 2001).

Conclusion

Energy storage is one of several potentially important enabling technologies supporting large- scale deployment of renewable energy, particularly variable renewables such as solar PV and wind. Energy storage is used in electric grids in the United States and worldwide. It is dominated by PSH. In addition to PSH, high-energy batteries and CAES can provide energy management services—shifting energy from periods of low demand to periods of high demand, which reduces curtailment and eases integration challenges associated with high levels of variable renewable generation—and were included in the RE Futures analysis. New storage capacity was deployed in all of the modeled scenarios and greater storage deployment is realized in scenarios with greater levels of renewables, and particularly variable renewable, penetration.

Capital cost is a primary barrier to deployment of energy storage. In addition, many regulatory and market barriers prevent energy storage from competing equally with more conventional technologies that provide energy and capacity services. A key issue for large-scale deployment of new storage capacity is finding suitable geologic formations for conventional PSH and CAES. PSH also faces potential opposition due to environmental impacts, which can be partially mitigated using closed-cycle plants. Both PSH and CAES plants are typically large, requiring new high-voltage transmission, which adds additional challenges, especially considering potentially remote locations. Batteries do not have the geologic constraints of CAES or PSH but large-scale deployment may face challenges related to a combination of materials requirements and competition with automotive applications.

More comprehensive estimates of the cost and resource availability for both CAES and PSH, advances in batteries to reduce capital cost, increase power and energy density, and increase lifetimes, and changes in market and regulations to quantify and value the ancillary services provided by energy storage are needed to support large-scale deployment of energy storage technologies in a high renewable electricity future.

References

ADM (2006). *Evaluation of Demonstration Project for Ice Bear® Thermal Ice Storage System for Demand Shifting*. Prepared by ADM Associates for Sacramento Municipal Utility District (SMUD). Sacramento, CA: SMUD.

Allen, A. E. (1977). "Potential for Conventional and Underground Pumped-Storage." *IEEE Transactions on Power Apparatus and Systems* (PAS-96, *3*), 993–998.

Allen, K. (1985). "CAES: The Underground Portion." *IEEE Transactions on Power Apparatus and Systems* (PAS-104, 4), 809–812.

Ali, M. H., Wu, B. & Dougal, R. A. (2010). "An Overview of SMES Applications in Power and Energy ystems." *IEEE Transactions on Sustainable Energy* (1, 1), 38–47.

APS (American Physical Society). (2007). "Challenges of Electricity Storage Technologies: A Report from the APS Panel on Public Affairs Committee on Energy and Environment." http://www.aps.org/policy/reports/popa-reports/upload/Energy-2007-Report-ElectricityStorageReport.pdf. Accessed February 18, 2012.

ASCE (American Society of Civil Engineers). Task Committee on Pumped Storage. (1993). *Compendium of Pumped Storage Plants in the United States*. New York: ASCE.

Baker, J. (2008). "New Technology and Possible Advances in Energy Storage." *Energy Policy*, (*36*), 4368–4373.

Bauer, S. J. & Webb, S. W. (2000). "*Summary Report on Studies and Analyses Supporting Underground Aspects of a CAES Facility at Norton, Ohio.*" SAND2000–3111. Albuquerque, NM: Sandia National Laboratories.

Black & Veatch. (2005). "*Iowa Stored Energy Plant: Economic Feasibility Analysis*." Project 139146. Overland Park, KS: Black & Veatch Corporation.

Black & Veatch. (2012). *Cost and Performance Data for Power Generation Technologies*. Overland Park, KS: Black & Veatch Corporation.

CEC (California Energy Commission). (2008). *Compressed Air Energy Storage [CAES] Scoping Study for California.* Prepared by Electric Power Research Institute for California Energy Commission Public Interest Energy Research Program. Report CEC–500–2008–069.

Sacramento, CA: California Energy Commission. http://www.energy.ca.gov/2008publications/CEC-500-2008-069/CEC-500-2008-069.PDF. Accessed February 18, 2012.

Clugston, J. P., ed. (1980). *Proceedings of the Clemson Workshop on Environmental Impacts of Pumped Storage Hydroelectric Operations.* Clemson, SC: U.S. Department of the Interior Fish and Wildlife Service Southeast Reservoir Investigations.

Crotogino, F., Mohmeyer, K.-U. & Scharf, R. (2001). "Huntorf CAES: More Than 20 Years of Successful Operation." Solution Mining Research Institute.

Dames & Moore. (1981). *An Assessment of Hydroelectric Pumped Storage.* Report IWR 82-H-10 prepared under contract to U.S. Army Engineer Institute for Water Resources. Washington, DC: Dames and Moore.

Deane, J. P., Ó Gallachóir, B. P. & McKeogh, E. J. (2010). "Techno-Economic Review of Existing and New Pumped Hydro Energy Storage Plant." *Renewable and Sustainable Energy Reviews* (14, 4), 1293–1302.

Denholm, P. & Holloway, T. (2005). "Improved Accounting of Emissions from Utility Energy Storage System Operation." *Environmental Science and Technology* (39, 23), 9016–9022.

Denholm, P. & Kulcinski, G. L. (2004). "Life Cycle Energy Requirements and Greenhouse Gas Emissions from Large Scale Energy Storage Systems." *Energy Conversion and Management* (45/13–14), 2153–2172.

Denholm, P. & Short, W. (2006). "An Evaluation of Utility System Impacts and Benefits of Optimally Dispatched Plug-In Hybrid Electric Vehicles." NREL/TP-620-40293. Golden, CO: National Renewable Energy Laboratory.

Denholm, P. & Sioshansi, R. (2009). "The Value of Compressed Air Energy Storage with Wind in Transmission-Constrained Electric Power Systems." *Energy Policy*, (37, 8), 3149–3158.

Denholm, P., Ela, E., Kirby, B. & Milligan, M. (2010). "*The Role of Energy Storage with Renewable Electricity Generation.*" NREL/TP-6A2-47187. Golden, CO: National Renewable Energy Laboratory. http://www.nrel.gov/docs/fy10osti/47187.pdf.

Desai, N., Nelson, S., Garza, S., Pemberton, D. J., Lewis, D., Reid, W., Lacasse, S., Spencer, R., Manning, L. M. & Wilson, R. (2003). "Study of Electric Transmission in Conjunction with Energy Storage Technology." Lower Colorado River Authority, Texas State Energy Conservation Office.

DOE (U.S. Department of Energy). (2010a). "Batteries for Electrical Energy Storage in Transportation (BEEST)." DE-FOA-0000207, Modification 003, CFDA 81.135.

DOE. (2010b). "Grid-Scale Rampable Intermittent Dispatchable Storage (GRIDS)." DE-FOA-0000290, CFDA 81.135. https://arpa-e-foa.energy.gov/FoaDetailsView.aspx?foaId=85e239bb-8908-4d2c-ab10-dd02d85e7d78. Accessed February 21, 2012.

DOE. (2007). *Basic Research Needs for Electrical Energy Storage.* Report of the Basic Energy Sciences (BES) Workshop on Electrical Energy Storage, April 2–4. http://science.energy.gov/~/media/bes/pdf/reports/files/ees_rpt.pdf. Accessed February 21, 2012.

EAC (Electricity Advisory Committee). (2008). *Bottling Electricity: Storage as a Strategic Tool for Managing Variability and Capacity Concerns in the Modern Grid.* Washington, DC: Office of Electricity Delivery and Energy Reliability of U.S. DOE.

EIA (U.S. Energy Information Administration). (2008). "Electricity Generating Capacity: Existing Electric Generating Units in the United States, 2008."

Washington, DC: U.S. DOE. http://www.eia.doe.gov/cneaf/electricity/page/capacity/capacity.html.

EIA. (2010). *Updated Capital Cost Estimates for Electricity Generation Plants.* Washington, DC: U.S. EIA Office of Energy Analysis, U.S. DOE. http://www.eia.gov/oiaf/beck_plantcosts/ pdf/updatedplantcosts.pdf.

EPRI (Electric Power Research Institute). (2010). *Electricity Energy Storage Technology Options: A White Paper Primer on Applications, Costs, and Benefits.* Report 1020676. Palo Alto, CA: EPRI.

EPRI, DOE. (2003). *EPRI-DOE Handbook of Energy Storage for Transmission and Distribution Applications.* Report 1001834. Palo Alto, CA: EPRI, Washington, DC: DOE.

EPRI, DOE. (2004). *Energy Storage for Grid Connected Wind Generation Applications. EPRI- DOE Handbook Supplement.* Report 1008703. Palo Alto, CA: EPRI, Washington, DC: DOE.

ESA (Electricity Storage Association). (2011). http://www.electricitystorage.org/images/uploads/static_content/technology/technolog y_resource s/ratings_large.gif. Accessed June 6, 2012.

ESA. (2011) "*Pumped Hydro.*" http://www.electricitystorage.org/technology/storage_technologies/pumped hydro/. Accessed February 18, 2012.

Eyer, J. & Corey, G. (2010). *Energy Storage for the Electricity Grid: Benefits and Market Potential Assessment Guide. A Study for the DOE Energy Storage Systems Program.* SAND2010-0815. Albuquerque, NM: Sandia National Laboratories.

Fagnard, J.-F., Crate, D., Jamoye, J.-F., Laurent, Ph., Mattivi, B., Cloots, R., Ausloos, M., Genon, A. & Vanderbemden, Ph. (2006). "Use of a High-Temperature Superconducting Coil for Magnetic Energy Storage." *Journal of Physics: Conference Series*, (*43*), 829–832.

Feak, S. (1997). "Superconducting Magnetic Energy Storage (SMES) Utility Application Studies." *IEEE Transactions on Power Systems*, (12, 3), 1094–1102.

FERC (Federal Energy Regulatory Commission). (n.d.). "All Issued Preliminary Permits." http://www.ferc.gov/industries/hydropower/gen-info/licensing/issued-pre-permits.xls. Accessed December 2011.

Gagnon, L. & van de Vate, J. F. (1997). "Greenhouse Gas Emissions from Hydropower. The State of Research in 1996." *Energy Policy*, (25, 1), 7–13.

Grazzini, G. & Milazzo, A. (2008). "Thermodynamic Analysis of CAES/TES Systems for Renewable Energy Plants." *Renewable Energy*, (33, 9), 1998–2006.

Greene, D. L., Leiby, P. N., James, B., Perez, J., Melendez, M., Milbrandt, A., Unnasch, S. & Hooks, M. (2008). *Hydrogen Scenario Analysis Summary Report: Analysis of the Transition to Hydrogen Fuel Cell Vehicles and the Potential Hydrogen Energy Infrastructure Requirements.* ORNL/TM-2008/030. Oak Ridge, TN: Oak Ridge National Laboratory. http://info.ornl.gov/sites/publications/files/Pub10268.pdf.

Grimsrud, G. P., Lefton, S. A. & Besuner, P. M. (2003). "True Cost of Cycling Power Plants Enhances the Value of Compressed Air Energy Storage (CAES) Systems." Presented at. *EESAT 2003 Electrical Energy Storage—Applications and Technology Conference*, San Francisco, October 2003.

Hadjipaschalis, I., Poullikkas, A. & Efthimiou, V. (2009). "Overview of Current and Future Energy Storage Technologies for Electric Power Applications." *Renewable and Sustainable Energy Reviews*, (*13*), 1513–1522.

Hall, P. J. & Bain, E. J. (2008). "Energy-Storage Technologies and Electricity Generation." *Energy Policy*, (36, 12), 4352–4355.

Hauch, A., Jensen, S. H., Ramousse, S. & Mogensen, M. (2006). "Performance and Durability of Solid Oxide Electrolysis Cells." *Journal of the Electrochemical Society*, (153, 9), A1741– A1747.

Hobson, M. J., Heath, E. G., Giramonti, A. J. & Adent, W. A. (1977). "Feasibility of CAES in California." Report to California Energy Commission. DOE/SF/90371-T1.

Ingram, E. A. (2010). "Worldwide Pumped Storage Activity." *Hydro Review Worldwide* (18, 4). Kaplan, S.M. (2009). *Electric Power Storage.* Congressional Research Service (CRS) Report R40797. http://assets.opencrs.com/rpts/R40797_20090908.pdf.

KEMA. (2008). "Summary of KEMA Validation Report: Two Megawatt Advanced Lithium-ion BESS Successfully Demonstrates Potential for Utility Applications." http://www.b2i.cc/ Document/546/KEMA_Report.pdf. Accessed July 2010.

KEMA. (2010). *Research Evaluation of Wind Generation, Solar Generation, and Storage Impact on the California Grid.* Prepared by KEMA for California Energy Commission Public Interest Energy Research Program. Report CEC-500-2010-010. http://www.energy.ca.gov/2010publications/CEC-500-2010-010/CEC-500-2010-010.PDF.

Kepshire, D. (2010). "Isothermal Compressed Air Energy Storage." *2010 Energy Storage Systems Research Program Update Conference*, Washington, D.C., 2010. http://www.sandia. gov/ess/docs/pr_conferences/2010/kepshire_sustainx.pdf.

Koh, H. & Magee, C. L. (2008). "A Functional Approach for Studying Technological Progress: Extension to Energy Technology." *Technological Forecasting and Social Change*, (*75*), 735–758.

Korinek, K., Clark, P. & Swensen, E. (1991). "Geological Screening for Compressed Air Energy Storage Plants." In *Proceedings of the American Power Conference*, Chicago, 1991.

Kroposki, B., Levene, J., Harrison, K., Sen, P. K. & Novachek, F. (2006). "Electrolysis: Information and Opportunities for Electric Power Utilities." NREL/TP-581-40605. Golden, CO: National Renewable Energy Laboratory. http://www.nrel.gov/docs/fy06osti/40605.pdf.

Luongo, C. A. (1996). "Superconducting Storage Systems: An Overview." *IEEE Transactions on Magnetics*, (32, 4), 2214–2223.

MacCracken, M. M. (2009). "Thermal Energy Storage Is Electric Energy Storage." Presented at *EESAT 2009 Electrical Energy Storage Applications and Technology Conference*, Seattle, WA.

Makarov, Y. V., Ma, J., Lu, S. & Nguyen, T. B. (2008). *Assessing the Value of Regulation Resources Based on Their Time Response Characteristics*. PNNL-17632. Richland, WA: Pacific Northwest National Laboratory. http://www.pnl.gov/main/publications/external/technical reports/PNNL-17632.pdf.

Mehta, B. (1992). "CAES Geology." *EPRI Journal*, (*17*, 7), 38–41.

Milliken, C. & Ruhl, R. (2003). "Low Cost, High Efficiency Reversible Fuel Cell Systems." NREL/CP-610-32405. FY 2003 Annual Progress Report for Hydrogen, Fuel Cells, and Infrastructure Technologies Program. DOE Hydrogen and Fuel Cells Annual Merit Review. In *Proceedings of the 2002 U.S. DOE Hydrogen Program Review*, Golden, CO.

Nakhamkin, M. (2008). "Second Generation of the CAES Technology." Presented at *Compressed Air Energy Storage (CAES) Scoping Workshop*, Center for Life Cycle Analysis, Columbia University, New York, October 21–22.

NGK. (n.d.). (2011). "Principle of the NAS Battery." NGK Insulators, Ltd. Accessed October 10. http://www.ngk.co.jp/english/products/power/nas/principle/.

Norton Energy Storage. (2000). "Application to the Ohio Power Siting Board for Construction of a Compressed Air Energy Storage Facility." Summit County, OH: Norton Energy Storage.

Nourai, A. (2007). *Installation of the First Distributed Energy Storage System (DESS) at American Electric Power (AEP): A Study for the DOE Energy*

Storage Systems Program. SAND2007-3580. Albuquerque, NM: Sandia National Laboratories.

NWPCC (Northwest Power and Conservation Council). (2008). Presented to *Pumped Hydro Storage Workshop*, Portland, OR, October 17, 2008. http://www.nwcouncil.org/energy/wind/meetings/2008/10/. Accessed February 20, 2012.

NYSEG (New York State Electric and Gas). (2009). *Compressed Air Energy Storage Engineering and Economic Study*. Final Report 10-09. Prepared for New York State Energy Research and Development Authority by New York State Electric and Gas (NYSEG). Albany, NY: NYSERDA. http://www.nyserda.ny.gov/en/Publications/Research-and-Development/~/media/Files/Publications/Research/Electic%20Power%20Delivery/10-09-compress-air-energy-storage.ashx. Accessed February 20, 2012.

ORNL (Oak Ridge National Laboratory): NHA (National Hydropower Association); HRF (Hydropower Research Foundation). (2010). "Pumped Storage Hydropower." Summary report on a summit meeting convened by ORNL, NHA, and HRF, September 20–21, 2010, Washington, DC. http://www.esd.ornl.gov/WindWaterPower/PumpedStorageSummitSummarySep2010.pdf.

Ohio Power Siting Board. (2001). "In the Matter of the Application of Norton Energy Storage, LLC for a Certificate of Environmental Compatibility and Public Need for an Electric Power Generating Facility in Norton, Ohio." Case 99-1626-EL-BGN.

Parfomak, P. W. (2012). *Energy Storage for Power Grids and Electric Transportation: A Technology Assessment*. R42455. Washington, D.C.: Congressional Research Service.

Peterson, S. B., Apt, J. & Whitacre, J. F. (2010). "Lithium-Ion Battery Cell Degradation Resulting from Realistic Vehicle and Vehicle-to-Grid Utilization." *Journal of Power Sources*, (195, 8), 2385–2392.

Peterson, S. B., Whitacre, J. F. & Apt, J. (2010). "The Economics of Using PHEV Battery Packs for Grid Storage." *Journal of Power Sources*, (195, 8), 2377–2384.

Phillips, J. (2000). "Pumped Storage in a Deregulated Environment." *International Journal on Hydropower and Dams*, (7, 1), 32–35.

Rastler, D. (2008). "New Demand for Energy Storage." *Electric Perspectives* (September/October [33, 5]) http://www.eei.org/magazine/EEI Electric Perspectives Article Listing/2008-09-01-EnergyStorage.pdf. Accessed December 2009.

Rastler, D. (2009). "*Overview of Electric Energy Storage Options for the Electric Enterprise.*" Palo Alto, CA: Electric Power Research Institute.

Rettberg, J. (2010). "Seneca Advanced Compressed Air Energy Storage (CAES) 150 MW Plant Using an Existing Salt Cavern." New York: New York State Electric and Gas. http://www.sandia.gov/ess/docs/pr_conferences/2010/rettberg_nyseg.pdf.

Rogers, L., Key, T. March, P. (2010). "Quantifying the Value of Hydropower in the Electric Grid." Presentated at *4th International Conference on Integration of Renewable and Distributed Energy Resources*, December 6–10, Albuquerque, NM. http://www.4thintegrationconference.com/downloads/ 7.04.pdf.

Rosa, L. P. & dos Santos, M. A. (2000). *Certainty and Uncertainty in the Science of Greenhouse Gas Emissions from Hydroelectric Reservoirs (Part 2): Thematic Review*. Cape Town, South Africa: World Commission on Dams.

RWE. (2010). "ADELE – Adiabatic Compressed-Air Energy Storage for Electricity Supply." Cologne, Germany: RWE Power AG. http://www.rwe.com/web/cms/mediablob/en/391748/ data/364260/1/rwe-power-ag/innovations/adele/Brochure-ADELE.pdf.

SDCWA (San Diego County Water Authority). (2011). "Lake Hodges Projects." http://www.sdcwa.org/sites/default/files/files/publications/lakehodges-fs.pdf. Accessed February 20, 2012.

Schalge, R. & Mehta, B. (1993). "The Alabama Electric Compressed Air Storage Cavern from Planning to Completion." In *Proceedings of the 55th American Power Conference*, Chicago, IL.

Schulte, R. H., Critelli, N., Jr. Holst, K. & Huff, G. (2012). *Lessons from Iowa: Development of a 270 Megawatt Compressed Air Energy Storage Project in Midwest Independent System Operator. A Study for the DOE Energy Storage Systems Program.* SAND2012-0388. Albuquerque, NM: Sandia National Laboratories.

Sioshanshi, R. & Denholm, P. (2010). "The Value of Plug-In Hybrid Electric Vehicles as Grid Resources" *The Energy Journal*, (31, 3), 1–23.

Smith, T. (2008). "Opportunities for Subsurface Compressed Air Energy Storage in New York State." Presented at *Compressed Air Energy Storage (CAES) Scoping Workshop*, Center for Life Cycle Analysis, Columbia University, New York, October 21–22.

Southeastern Power Administration. (2009). "Hydropower Pump-back Projects/Perspectives." Presentation to *Southwestern Federal Hydropower*

Conference, June 10. http://www.swpa.gov/PDFs/2009Conference/Pumpback-PMA-Perspective-Nadler.pdf.

Sterner, M. (2009). *Bioenergy and Renewable Power Methane in Integrated 100% Renewable Energy Systems: Limiting Global Warming by Transforming Energy Systems*. Kassel, Germany: University of Kassel Press.

Steward, D., Saur, G., Penev, M. & Ramsden, T. (2009). *Lifecycle Cost Analysis of Hydrogen Versus Other Technologies for Electrical Energy Storage*. NREL/TP-560-46719. Golden, CO: National Renewable Energy Laboratory. http://www.nrel.gov/docs/fy10osti/46719.pdf.

Strauss, P. L. (1991). "Pumped Storage, the Environment and Mitigation." In *Proceedings of Waterpower '91: A New View of Hydro Resources*, Denver, CO, July 24–26. New York: American Society of Civil Engineers.

Succar, S. & Williams, R. D. (2008). *Compressed Air Energy Storage: Theory, Operation and Applications*. Princeton Environmental Institute, Princeton University.

Suresh, B., Schlag, S., Kumamoto, T. & Ping, Y. (2010). Hydrogen, Chemical Economics Handbook (CEH) Marketing Research Report, SRI Consulting.

Tam, G. (2008). "Eagle Mountain Hydro-Electric Pumped Storage Project," Eagle Crest Energy Company presentation to Northwest Wind Integration Forum, Portland, OR, October 17.

Tanaka, H. (2000). "The Role of Pumped-Storage in the 21st Century." *International Journal on Hydropower and Dams* (*7*, 1), p. 27.

Thoms, R. L. & Gehle, R. M. (2000). "A Brief History of Salt Cavern Use." Presented to *8th World Salt Symposium*, The Hague, Netherlands, May 7–11.

TIAX. (2002). *Grid-Independent, Residential Fuel-Cell Conceptual Design and Cost Estimate*. Reference 76570. Final Report for DOE National Energy Technology Laboratory. Cambridge, MA.

TMI (Technology Management Inc.). (2001). "Low Cost, High Efficiency Reversible Fuel Cell (and Electrolyzer) Systems." In *Proceedings of the 2001 DOE Hydrogen Program Review*. NREL/CP-570-30535. Cleveland, OH: TMI. http://www1.eere.energy.gov/hydrogenandfuelcells/pdfs/30535aw.pdf.

Troy, N., Denny, E. & O'Malley, M. (2010). "Base-Load Cycling on a System with Significant Wind Penetration." *IEEE Transactions on Power Systems* (25, 2), 1088–1097.

TVA (Tennessee Valley Authority). (2001). "Environmental Assessment: The Regenesys™ Energy Storage System." Muscle Shoals, AL: TVA. http://www.tva.gov/environment/reports/regenesys/preface.pdf.

U.S. Bureau of Reclamation, Colorado State University (CSU) Cooperative Fishery Unit. (1993). *Aquatic Ecology Studies of Twin Lakes, Colorado 1971–86: Effects of a Pumped-Storage Hydroelectric Project on a Pair of Montane Lakes*. Denver, CO: U.S. Bureau of Reclamation.

U.S. GAO (General Accounting Office). (1996). *Power Marketing Administrations: Cost Recovery, Financing, and Comparison to Nonfederal Utilities*. GAO/AIMD-96-145. Washington, D.C.: U.S. GAO. http://www.gao.gov/archive/1996/ai96145.pdf.

Wadia, C., Albertus, P. & Srinivasan, V. (2011). "Resource Constraints on the Battery Energy Storage Potential for Grid and Transportation Applications." *Journal of Power Sources*, (*196*), 1593–1598.

Walawalkar, R., Apt, J. & Mancini, R. (2007). "Economics of Electric Energy Storage for Energy Arbitrage and Regulation in New York." *Energy Policy*, (35, 4), 2558–2568.

Willis, R. & Parsonnet, B. (2010). *Energy Efficient TES Designs for Commercial DX Systems*. OR-10-016. Atlanta, GA: American Society of Heating, Refrigerating and Air-Conditioning Engineers.

Yang, C.-J. & Jackson, R. B. (2011). "Opportunities and Barriers to Pumped-Hydro Energy Storage in the United States." *Renewable and Sustainable Energy Reviews*, (*15*), 839–844.

Yang, Z., Zhang, J., Kintner-Meyer, M. C. W., Lu, X., Choi, D., Lemmon, J. P. & Liu, J. (2011). "Electrochemical Energy Storage for Green Grid." *Chemical Reviews*, (111, 5), 3577–3613.

Yasuda, M. (2000). "Enhancing Ancillary Services to Make Pumped Storage More Competitive." *International Journal on Hydropower and Dams*, (7, 1), 36–42.

End Notes

[164] Estimates for the total installed capacity for PSH in the United States range from 20 GW to 22 GW. This range is partially due to the use of different plant ratings. For example, the EIA lists the total nameplate capacity of PSH as of 2008 at 20.4 GW, while the summer capacity is listed at 21.9 GW.

[165] Areas in the United States with wholesale energy markets typically also include markets for both spinning contingency reserves and regulation reserves.

[166] However, in the ReEDS and GridView modeling, storage devices also contribute to ancillary services (e.g., forecast error, contingency, and frequency regulation reserves).

[167] The compressed air can be considered a method to assist conventional natural gas turbines by providing the compressed air that typically requires about two thirds of the energy generated by a gas turbine. This reduces the natural gas fuel used by a gas turbine by more than 50%, reducing the heat rate from approximately 10,000 Btu/kWh to approximately 4,000 Btu/kWh (Succar and Williams 2008).

[168] This conclusion depends on the anticipated cycle life and cost of EV batteries. See Sioshansi and Denholm (2010) and Peterson, Whitacre, and Apt (2010) for a discussion of the impact of battery life and cycling on the value of V2G. However, controlled charging (without V2G) is still a potentially significant source of flexibility, with the ability to raise the minimum load and avoid curtailment.

[169] A full cell is a device capable of generating an electrical current by converting the chemical energy of a fuel (e.g., hydrogen) directly into electrical energy. Fuel cells differ from conventional electrical (e.g., battery) cells in that the active materials such as fuel and oxygen are not contained within the cell but are supplied from outside. It does not contain an intermediate heat cycle, as do most other electrical generation techniques (http://www.eia.doe.gov/glossary/).

[170] There is one small PSH facility under construction as of November 2011 (the 40-MW Olivenhain-Hodges project) with completion expected in 2012 (SDCWA 2011).

[171] All dollar amounts presented in this report are presented in 2009 dollars unless noted otherwise; all dollar amounts presented in this report are presented in U.S. dollars unless otherwise noted.

[172] The ARPA-E goal of $100/kWh includes both the power and energy component, including power conditioning equipment, installation, and other balance of system components. This corresponds to $800/kW for a device with 8 hours of storage capacity. This would require battery costs of well below $100/kWh, considering balance of system is currently a considerable fraction of $800/kW (U.S. DOE 2010b).

[173] See Volume 1 for details on the design of the scenarios.

[174] As a cost optimization model, ReEDS produces deployment results that can fluctuate greatly from year to year, whereas the actual deployment of technologies tends to vary more smoothly over time.

[175] While ancillary services markets have been created in locations with restructured markets, large areas of the United States, including the entire West (excluding California) and most of the Southeast.

INDEX

D

E

F

G

N

O

P

R

S

W

Y

Z